Environmental Science and Engineering

The ultimate goal of this series is to contribute to the protection of our environment, which calls for both profound research and the ongoing development of solutions and measurements by experts in the field. Accordingly, the series promotes not only a deeper understanding of environmental processes and the evaluation of management strategies, but also design and technology aimed at improving environmental quality. Books focusing on the former are published in the subseries Environmental Science, those focusing on the latter in the subseries Environmental Engineering.

fit the negative stereotype are considered old, and those who defy the stereotype are considered "still young." In order to protect ones aging social identity, people may maintain the negative stereotype and dissociate themselves from it. This does not mean that they view aging or old age in a more positive way. It means that they consider themselves immune in face of the "threat" of growing old themselves. The divide between mind and body in the process of aging is the effect of a youth obsessed culture. For middle-aged individuals living in precarious contexts, the body–mind dualization may represent on the one hand a growing divide between those who are old and those who are "still young." On the other hand, the body–mind dualization stands for aging denial and signify a divide between the internal and external self. One way to solve this issue is to try to integrate the body and mind in the aging process. This could maybe be achieved by adopting alternative positive images of aging.

When individuals deny aging or old age, they do not reject the negative aging stereotype (Andrews 1999). They rather reject the fact that they could be old or become old themselves. Andrews (1999) argues that being told that one "does not look old" is taken as a compliment, but in fact it just means one erases the years of experiences one has had. Old people are not young, they have lived a lot and developed over the years. Stating that "one does not appear old" represents a denial of all the changes, especially psychological ones, that a person has gone through (good and bad). This situation begs the question "Why is aging not seen as an accomplishment in itself?" The denial of aging represents a negative image of old age and stands for the fears they have about growing old. Especially for precarious individuals, negative images of old age are expressed as "no representation of aging." Precarious people prefer not to think about old age and live in the present. The "precarious mind" as Guy Standing (2011) has called it is plagued by anxieties and by short-termism. Rather than dreaming about travels or other leisure pursuits in their retirement years, precarious persons are worried about absent pensions and lack of money to pay the rent or insurance. Another worry concerns that of not being fit enough to work, or not being eligible for work because one is "too old" for the job.

To summarize, using the body as "an identity card" and considering chronological age unimportant support the anti-aging philosophy. Life stages are constructed socially and culturally. When chronological anchors (e.g., retirement age) are absent, it becomes difficult to define and delimitate old age from other life stages. Instead of appealing to external social conventions such as the age of entering retirement, physical signs, and symptoms become salient when describing the onset of old age. Precarity, understood here as the retreat of the institutions of the welfare state that would help define old age, makes the whole concept of old age fluid or indefinite. Thus, the concept of old age becomes precarious.

Although both categories of participants agree that the cliché image of the "old frail person" is changing, the precarious middle-aged participants were more concerned with the topic of "staying fit as long as possible." Even though it may be tempting to be happy about the fact that people want to look younger or that they "feel younger" than they actually are, such thoughts are based on negative views of old age. Desiring

Darren Jones · Ahmed Jalal Khan Chowdhury
Editors

Environmental Resilience in the Pandemic Years 2020–2021

COVID-19 and Environmental Ecosystem

Springer

Editors
Darren Jones
INWASCON
International Water Air and Soil Conservation Society
Kuala Lumpur, Malaysia

Ahmed Jalal Khan Chowdhury
Faculty of Agriculture
Universiti Islam Sultan Sharif Ali (UNISSA)
Tutong, Brunei Darussalam

ISSN 1863-5520 ISSN 1863-5539 (electronic)
Environmental Science and Engineering
ISBN 978-3-031-47756-0 ISBN 978-3-031-47757-7 (eBook)
https://doi.org/10.1007/978-3-031-47757-7

This Springer imprint is published by the registered company Springer Nature Switzerland AG
The registered company address is: Gewerbestrasse 11, 6330 Cham, Switzerland

Contents

Chapter 1
Covid-19 Pandemic: Towards the Nature's Ecological Resilience

Musa Najiah, Kok Leong Lee, Musa Nadirah, Rasina Rasid, and Hassan Ibrahim Sheikh

Abstract The novel coronavirus COVID-19 pandemic or known as COVID-19 pandemic has triggered a global public health crisis. Drastic and sweeping measures taken by countries around the world to curb the spread of COVID-19, including border shutdowns, travel bans, full or partial lockdowns, business closures, home confinement, and others have brought economic activities to a near halt, leading to an economic crisis. These measures have also had positive and negative effects on the environment and biodiversity. Nevertheless, the pandemic has had significant effects on nature's ecological resilience. Our review demonstrates a significant association between contingency measures and improvements in air and land quality, as well as an increase in wildlife populations. However, there have also been negative effects such as an increase in waste, which further endangers the water and land ecology. Ecological resilience is essential since all ecosystems, whether lithosphere, atmosphere or hydrosphere, face hazards that require the identification of mitigation risks. Resilience can be seen as a factor influencing adaptive capacity, which, in turn, is an element of the vulnerability of nature's ecological system. This paper provides a brief review of nature's ecological resilience and sustainability during the COVID-19 pandemic.

M. Najiah (✉) · K. L. Lee · M. Nadirah · R. Rasid · H. I. Sheikh
Faculty of Fisheries and Food Science, Universiti Malaysia Terengganu (UMT), 21030 Kuala Nerus, Terengganu, Malaysia
e-mail: najiah@umt.edu.my

K. L. Lee
e-mail: drleekl@gmail.com

M. Nadirah
e-mail: nadirah@umt.edu.my

R. Rasid
e-mail: rrasina@umt.edu.my

H. I. Sheikh
e-mail: sheikh@umt.edu.my

D. Jones and A. J. Khan Chowdhury (eds.), *Environmental Resilience in the Pandemic Years 2020–2021*, Environmental Science and Engineering,
https://doi.org/10.1007/978-3-031-47757-7_1

Keywords SARS-COV-2 · Lockdown · Movement control · Reversibility and sustainability · Anthropogenic origin

1.1 Introduction

The new coronavirus disease that emerged at the end of 2019 quickly became a pandemic, triggering a worldwide public health and humanitarian crisis. The disease was later named coronavirus disease 2019 (COVID-19), and its etiological agent was identified as SARS-CoV-2. In an effort to contain this highly contagious disease and flatten the epidemic curve, social lockdowns were implemented by numerous countries and regions around the world in 2020, following the WHO's global response strategy and health recommendations (WHO 2020). These drastic and sweeping measures resulted in temporary halts to many aspects of daily life, including school and business closures, border shutdowns, travel bans, and home confinement. While the lockdown did not fully curb the disease, it did slow down the pace of the pandemic to some extent. COVID-19 has had a significant impact on every aspect of humanity, both positive and negative. As of 4 January 2023, there have been over 661 million confirmed cases of the disease worldwide, with approximately 6.7 million lives lost (Johns Hopkins University Coronavirus Resource Center 2021). This article reviews and discusses some aspects of nature's resilience as seen in COVID-19, including susceptibility and capability to face harsh conditions, which are key factors in resilience.

Ecological resilience refers to an ecosystem's ability to maintain its normal patterns of nutrient cycling and biomass production after being subjected to damage caused by an ecological disturbance, whether in the lithosphere, atmosphere, or hydrosphere. Nature's ecological resilience involves various aspects and can be related to activities that occur on land and in water. For example, sectors related to soil biodiversity, forest resources, livestock, water, fisheries, aquaculture, and wildlife management have been among the most affected areas (FAO 2020a, b; Hockings et al. 2020; Lal et al. 2020; Poch et al. 2020). The direct influences of the COVID-19 pandemic on human activities depend on the situation in each country, such as a decline in human consumption leading to either a surplus or a reduction in food production.

1.2 Environment's Resilience

According to WHO (2022), air pollution is the world's single largest environmental health risk, in which almost 4.2 million premature deaths a year were reported due to outdoor pollution. It is estimated that 9 out of 10 people breathe air containing high levels of pollutants. As the majority of countries worldwide have implemented necessary protective measures to contain the disease, such as enforcing lockdowns and physical distancing measures, Earth's biomes are recuperating and healing. Reports have been surfacing worldwide of cleaner air with less carbon emissions. For instance, lockdowns help to reduce air pollution as many industrial operations have to stop, and fewer vehicles are on the roads, seas, and in the air. It has been widely reported that the lockdown period has led to improved air quality, especially in highly polluted industrial and residential areas. Furthermore, many reports indicate that improved air quality and reductions in key pollutants such as carbon oxides (CO_x), nitrogen oxides (NO_x), sulfur oxides (SO_x), and particulate matter (PM) emissions have lowered their concentrations in the atmosphere (Chen et al. 2020; Wang et al. 2020; Wang and Su 2020; Zambrano-Monserrate et al. 2020).

Air quality improvement has been related to quarantine and lockdown measures by comparing air quality in pre-lockdown, lockdown, relaxed lockdown, and post-lockdown periods. This has been observed in countries such as China (Chen et al. 2020; Pei et al. 2021; Yao et al. 2020), Brazil (Dantas et al. 2020; Nakada and Urban 2020), Egypt (Mostafa et al. 2021), India (Mahato et al. 2020), Italy (Fattorini and Regoli 2020), Spain (Baldasano 2020), the United States (Chen et al. 2020; Zangari et al. 2020), and many others worldwide, including Canada (Adams 2020) and Southeast Asia (Kanniah et al. 2020; Stratoulias and Nuthammachot 2020). Indeed, air quality improvement was the most beneficial environmental aspect due to the COVID-19 pandemic, mainly because of the reduced fuel consumption resulting from the shutdown of transportation and industrial sectors (Table 1.1). However, ozone levels in ambient air have been found to increase, perhaps in response to reduced nitrogen emissions. Nevertheless, improving air quality can help increase resilience against acute respiratory illnesses, as well as other lung and heart diseases, and premature death. As economies begin to recover from the pandemic, continued air, land, and sea travel, congested traffic, movement of people from one place to another, and increased production levels in factories are expected to lead to an increase in outdoor air pollution in the coming years.

Although the spread of SARS-CoV-2 is closely related to water and sanitation, fewer reports have discussed the improved water quality, mainly due to the shutdown of many industrial activities releasing wastewater into water bodies (Barceló 2020; Lokhandwala and Gautam 2020; Selvam et al. 2020a, b). On the other hand, some reports have indicated that wastewater has been loaded with additional and persistent organic matter due to the excessive use of sanitizers, disinfectants, medication, and antibiotics (Barcelo 2020; Espejo et al. 2020; Selvam et al. 2020a, b; Zambrano-Monserrate et al. 2020). For instance, Zuo (2020) reported that during the peak of the outbreak in Wuhan, approximately 240 tons of medical waste were generated daily compared to 40 tons daily before the outbreak. During the COVID-19 pandemic, water pollution has also been reduced in coastal areas of Bangladesh, Malaysia, Thailand, Maldives, Indonesia, Acapulco (Mexico), and Barcelona (Spain) (Rume and Islam 2020; Mousazadeh et al. 2021). Furthermore, improved water quality in natural water resources has been reported, as shown in Table 1.2. On the other hand, wastewater facilities have reported increasing levels of organic matter and chemicals due to the increased use of sanitizers, disinfectants, and antibiotics. The solid waste generated due to the COVID-19 pandemic was found to increase both qualitatively

Table 1.1 Summary for the observed pollution parameters due to the COVID-19 pandemic lockdown

Regions	Pollution parameters					References
	NO_2	SO_2	O_3	$PM_{2.5}$	PM_{10}	
Malaysia (Kuala Lumpur)	1/1–17/3, 2020: 12.1 ppb 18/3–21/4, 2020: 5.54 ppb	1/1–17/3, 2020: 0.89 ppb 18/3–21/4, 2020: 0.57 ppb	1/1–17/3, 2020: 18.1 ppb 18/3–21/4, 2020: 19.9 ppb	1/1–17/3, 2020: 18.6 $\mu g/m^3$ 18/3–21/4, 2020: 19.3 $\mu g/m^3$	1/1–17/3, 2020: 24.3 $\mu g/m^3$ 18/3–21/4, 2020: 24.2 $\mu g/m^3$	Othman and Latif (2021)
China (Wuhan)	2/2019: 1.5E+16 molecules/cm^2 2/2020: 2.5E+15 molecules/cm^2	2/2019: 3.9E+15 molecules/cm^2 2/2020: 1.1E+15 molecules/cm^2	n/a	n/a	n/a	Ghahremanloo et al. (2021)
South Korea	3/2019: 20.38 ± 6.63 ppb 3/2020: 16.22 ± 4.95 ppb	n/a	No change	3/2019: 37.37 ± 23.95 $\mu g/m^3$ 3/2020: 20.39 ± 6.31 $\mu g/m^3$	3/2019: 60.77 ± 31.05 $\mu g/m^3$ 3/2020: 39.16 ± 7.23 $\mu g/m^3$	Ju et al. (2021)
Japan (Tokyo)	2/2019: 9.8E+15 molecules/cm^2 2/2020: 7.9E+15 molecules/cm^2	2/2019: 2.7E+15 molecules/cm^2 2/2020: 9.4E+15 molecules/cm^2	n/a	n/a	n/a	Ghahremanloo et al. (2021)
India (Kolkata)	5/2019: 16.48 $\mu g/m^3$ 5/2020: 10.66 $\mu g/m^3$	5/2019: 6.82 $\mu g/m^3$ 5/2020: 2.54 $\mu g/m^3$	5/2019: 31.92 $\mu g/m^3$ 5/2020: 38.68 $\mu g/m^3$	5/2019: 34.81 $\mu g/m^3$ 5/2020: 16.86 $\mu g/m^3$	5/2019: 88.99 $\mu g/m^3$ 5/2020: 35.66 $\mu g/m^3$	Bera et al. (2020)
Iraq	1/3–16/3,2020: 42 $\mu g/m^3$ 17/3–21/4, 2020: 39 $\mu g/m^3$	n/a	1/3–16/3, 2020: 40 $\mu g/m^3$ 17/3–21/4, 2020: 44 $\mu g/m^3$	1/3–16/3, 2020: 38 $\mu g/m^3$ 17/3–21/4, 2020: 29 $\mu g/m^3$	1/3–16/3, 2020: 132 $\mu g/m^3$ 17/3–21/4, 2020: 112 $\mu g/m^3$	Hashim et al. (2021)

(continued)

Table 1.1 (continued)

Regions	Pollution parameters					References
	NO_2	SO_2	O_3	$PM_{2.5}$	PM_{10}	
France (Lyon)	2/2020: 36.8 μg/m^3 17/3–11/5, 2020: 12.0 μg/m^3	n/a	2/2020: 33.5 μg/m^3 17/3–11/5, 2020: 68.9 μg/m^3	2/2020: 12.1 μg/m^3 17/3–11/5, 2020: 18.5 μg/m^3	2/2020: 20.0 μg/m^3 17/3–11/5, 2020: 24.5 μg/m^3	Sbai et al. (2021)
UK	2013–2019: 22.92 μm/m^3 23/3–30/6, 2020: 13.21 μm/m^3	2013–2019: 2.26 μm/m^3 23/3–30/6, 2020: 3.95 μm/m^3	2013–2019: 59.38 μm/m^3 23/3–30/6, 2020: 66.03 μm/m^3	2013–2019: 11.17 μm/m^3 23/3–30/6, 2020: 9.14 μm/m^3	n/a	Higham et al. (2021)
Italy (Nice)	2017–2019: 34.0 ± 7.3 μg/m^3 Lockdown 2020: 12.5 ± 2.4 μg/m^3	n/a	2017–2019: 62.6 ± 2.1 μg/m^3 Lockdown 2020: 77.6 ± 1.3 μg/m^3	2017–2019: 12.7 ± 0.9 μg/m3 Lockdown 2020: 12.4 ± 1.0 μg/m^3	2017–2019: 25.7 ± 3.9 μg/m^3 Lockdown 2020: 24.1 ± 3.4 μg/m^3	Sicard et al. (2020)
Spain (Barcelona)	16/3–24/5, 2015–2019: 33.3 μg/m^3 16/3–24/5, 2020: 15.3 μg/m^3	16/3–24/5, 2015–2019: 2.5 μg/m^3 16/3–24/5, 2020: 1.9 μg/m^3	16/3–24/5, 2015–2019: 85.3 μg/m^3 16/3–24/5,2020: 86.5 μg/m^3	n/a	n/a	Querol et al. (2021)

and quantitatively. KWR (2020) indicated that SARS-CoV-2 screening at municipal wastewater plants can help trace the evolution of the pandemic and detect future outbreaks. Indeed, an early warning system based on screening and information sharing on SARS-CoV-2 in wastewater among the public is important for quarantine and target monitoring.

Al Huraimel et al. (2020) found that the presence of SARS-CoV-2 in wastewater is unlikely to be transmitted via wastewater because Coronaviruses are sensitive

Table 1.2 Summary of some reports on water categories and its water quality parameters investigated during COVID-19 pandemic (Manoiu et al. 2022)

Location of water quality reported	Water categories	Parameters involved	References
Spain, Netherlands, Australia, USA, France, Italy, Turkey, Ecuador, China, India	Wastewater	SARS-CoV-2 virus and/or its RNA	Al Huraimel et al. (2020) and Bhowmick et al. (2020)
India	Surface water (river, lake), groundwater	Suspended particulate matter (SPM); heavy metals	Casado-Aranda et al. (2021)
Italy, Hong Kong, Canada, Germany, and Austria	Surface water, wastewater	SARS-CoV-2 virus; medical and hygienic waste	Cheval et al. (2020)
Asian countries, Italy	Surface water, wastewater	Chemical pollutants; plastic waste; and suspended solids	Espejo et al. (2020)
India, China, Malaysia, and Morocco	Surface water (rivers and lakes), sea/ocean water, wastewater	Biodiversity; macroplastic and microplastic pollution; chlorophyll a (Chl-a); phyto-plankton; nitrogen; turbidity; pharmaceuticals and disinfectants; and SARS-CoV-2 virus	Yusoff et al. (2021)
Southeast Asia (Indonesia, Malaysia, Thailand, the Philippines, and Vietnam)	Surface water (rivers), sea/ocean water	TSS; DO; and plastic and medical waste	Praveena and Aris (2021)
India	Groundwater	Physico-chemical parameters (pH, TDS, electrical conductivity (EC); NO_3F, heavy metals; As, Se); bacterial pollution parameters (TC; fecal coliforms; and *E. coli*; fecal streptococci)	Selvam et al. (2020a, b)

to disinfectants and organic solvents, which kill them within 2 to 3 days in sewage. Additionally, they found that although the virus was detectable in sewage water before and sometimes after chlorine disinfection, there was no living SARS-CoV. Accidental contact with treated wastewater may pose a minor potential risk of infection, but it can be a potential route of infection if in contact with untreated wastewater in below-par sanitary areas.

The COVID-19 lockdown has also seen an increasing trend of dissolved oxygen and abrupt fall in biochemical oxygen demand, nitrate, fecal coliform, and total coliform during the lockdown period in Damodar River, India (Chakraborty et al. 2021), and Ganga River (Dutta et al. 2020), where significant improvement of water quality had occurred due to the lockdown, leading the water to be fit for drinking (Class A) in the upper segments and for outdoor bathing (Class B) in the middle and lower segments, possibly due to reduced waste disposal following suspension of industrial and business activities (Karunanidhi et al. 2021; Mukherjee et al. 2020; Yunus et al. 2020; Selvam et al. 2020a, b; Khan et al. 2021). According to Kundu (2020), water pollution was minimized in Malaysia, Indonesia, Thailand, Bangladesh, and the Maldives due to the COVID-19 lockdown. Pal et al. (2020) reported that the COVID-19 lockdown has also accelerated the growth of phytoplankton species in the brackish system and estuary due to stress elimination posed by pollution from point and non-point sources. Indeed, the COVID-19 pandemic has resulted in a positive ecosystem direction, as these microorganisms are the base of the aquatic food pyramid, which influences the food web.

1.3 Biodiversity's Resilience

Biodiversity is important from the perspectives of natural heritage conservation, both terrestrial and aquatic (Manenti et al. 2020; Corlett et al. 2020; Rutz et al. 2020), as well as economic activities such as tourism (Bakar and Rosbi 2020) and agriculture (Rawal et al. 2020). Biodiversity, which covers forest, wetland and mangrove ecosystems and their contents, plays an important role in economic activities and livelihoods. It has been reported that biodiversity is often undervalued. For instance, the natural capital used by primary production such as agriculture, forestry, fisheries, mining as well as some primary processing sectors such as cement, steel, pulp and paper was valued at USD 7.3 trillion in 2013 (Natural Capital Coalition 2016). The health and sustainability of our world relies on a balanced biodiversity consisting of a wide range of fauna and flora, as well as microorganisms. Mankind's activities are the direct cause of biodiversity loss, for instance, land use for large-scale food production and human settlement, as well as overexploitation such as overfishing and overharvesting. In addition, overexploitation of habitats and wildlife trade play a vital part in disease transmission. The COVID-19 lockdown has led to a reduction in the intensity of human activities, which has brought more positive impacts on biodiversity preservation and recovery.

The lockdown has seen an unusual appearance of wild animals in places that they would normally avoid, leading to the enthusiastic thought that "Nature is coming back," even though the species clearly do not represent native biodiversity (Manenti et al. 2020). According to Parrino et al. (2021), there has indeed been an increase in reports of alien animal species in unusual habitats. In Malaysia, the smooth-coated otter (*Lutrogale perspicillata*), which is listed globally as a vulnerable species, has been seen roaming around lake areas in the Klang Valley, Selangor, such as Perdana Botanical Garden, Taman Tasik Metropolitan Kepong, and Subang Ria Park, as well as in Penang, specifically Sungai Pinang (MalayMail 2020). These smooth-coated otters have been listed on the CITES (Convention on International Trade in Endangered Species) Appendix I (previously only Appendix II since 1977), strengthening their protection from international trade. As otters prefer unpolluted water with minimum human disturbance, their presence in the lake area suggests improved lake water quality and the absence of human disruption. On the other hand, wild elephants were also seen wandering around domesticated areas in Kenyir, Hulu Terengganu, and tapirs were found wandering in Jementah, Johor (Malaysia News Yahoo 2020). Indeed, the lack of road and outdoor activities due to the movement control order (MCO) has led to wildlife encounters on the streets of Malaysia, with animals roaming contentedly to find food.

Conversely, the postponement of alien species eradication in some countries due to lockdown has led to the proliferation of various alien species, such as coypu (*Myocastor coypus*), pond slider (*Trachemys scripta*), wels catfish (*Silurus glanis*), and Eastern grey squirrel (*Sciurus carolinensis*) (Manenti et al. 2020). Furthermore, the invasive American mink (*Neovison vison*) has shown increased activity in Valdivia, Chile (Silva-Rodríguez et al. 2021). Biological invasion has a significant impact on island and freshwater ecosystems, which are some of the most vulnerable habitats (Sax and Gaines 2008).

As a result of COVID-19, the closure of fish landing sites, cold chain facilities, and fish markets has forced fishing fleets to discard much of their catches back into the sea. The suspension of fishing activities has somewhat eased fish stock depletion and had a positive impact on fisheries conservation. Highly manned and equipped foreign trawling vessels pose a severe threat to marine fish stocks and habitats. The suspension of recreational activities such as scuba diving and snorkeling has also had a positive impact on marine life conservation and recovery in the tourism industry.

Overall, effective biodiversity conservation can help reduce the risk of zoonotic transfer and maintain the existing wealth of ecosystem services. According to the OECD (2019), an ambitious and effectively implemented Post-2020 Global Biodiversity Framework under the United Nations Convention on Biological Diversity can serve as a venue for policy change that can help mitigate future disease outbreaks. Additionally, the importance of biodiversity should be better reflected in producer and consumer decision-making.

1.4 Hazard, Risk Mitigation and Way Forward

The term "risk" refers to the possibility of negative consequences in the future, including the chance of losing human lives, economic assets, environmental resources, cultural values, and critical infrastructure due to unexpected and destructive events (Adger et al. 2013). The National Research Council (NRC) defined "hazard" as a phenomenon that has the potential to cause harm or negative consequences to others, while "risk" refers to the possibility of something bad or unpleasant happening. Hazards are potential sources of danger or harm that can lead to risks, which are defined as the probability of adverse outcomes in the future, such as loss of life, damage to the environment, or economic consequences.

COVID-19 has prompted us to consider our relationship with the environment. Currently, global carbon dioxide emissions are expected to drop to levels seen a decade ago. However, as the world resumes normal activities, the increase in emissions may be greater than the decrease. Therefore, policies, economic incentives, and investments should focus on cleaner and more resilient energy infrastructure. For example, renewable energy and improved energy efficiency in all sectors can be implemented to address future challenges.

Confinement, physical distancing, epidemiological control, strengthened capacity of the public health system, fiscal and monetary measures have all contributed to surviving the COVID-19 pandemic. In addition, the enforcement of movement control order (MCO) has led to fewer illegal, unreported, and unregulated (IUU) activities, resulting in less destruction of natural resources. The COVID-19 pandemic crisis is one of the most significant unprecedented global events, teaching us the importance of slowing down. Drawing from the positive lessons learned from the COVID-19 pandemic, Table 1.3 presents recommended criteria for mitigating risks and facing adversity.

1.5 Conclusion

The reduction in anthropogenic activities during COVID-19 lockdown has provided a temporary demonstration of nature's resilience to pollution. Improvements in various environmental parameters suggest that the planet's self-regulatory mechanisms have been coping with pollution issues, likely since the industrial revolution. For durable and resilient economic recovery from the COVID-19 crisis, it is important to avoid returning to "business as usual" and environmentally destructive investment patterns and activities. By working together to strengthen laws and prioritize risk mitigation measures, we can support the ecological resilience of our natural world for many years to come.

Table 1.3 Recommended criteria for risk mitigation measures COVID-19 on environments and biodiversity

Categories	Hazard types	Risk mitigation measures	References
Human life	Human and environmental health	Energy efficient housing, building as part of compact, resilient and sustainable cities	I4CE (2020)
Animal life, ecosystem	Anthropogenic hazard/ecological hazard	Enhancing biodiversity to ensure a resilient supply of food, policies to value biodiversity, payments for ecosystem services, approaches to measure and value natural capital loss	OECD (2019, 2020)
Air pollution	Environmental health	Low-carbon, resilient electricity systems, systematic policies based on pollution sources and characteristics in each city with the use of cleaner alternative and new vehicle technology, reduced human outdoor activities, vehicle emissions and coal-fired power plant emissions	Othman and Latif 2021
Fisheries & Aquaculture	Ecological hazard	Proactive policies and regulatory frameworks, good management practices, good manufacturing practices, automation, digitization technology	Chowdhury et al. (2021) and Yusoff et al. (2021)
Food	Human health and food safety	National strategies on aquatic animal health, biosecurity, disease surveillance and reporting, early warning, emergency response and contingency planning, import risk analysis, good health management practices, vaccination, GIS risk mapping, consumer education, integrated approaches involving health education, vector control	Chowdhury et al. (2021) and Yusoff et al., (2021),
Alien Invasive Species	Ecological hazard	Maintaining strict screening protocols, implementation education campaigns about IAS, favoring the adoption of pathogens-contrasting protocols, awareness of the risks of pathogens spread	Parrino et al. (2021)

References

Adams MD (2020) Air pollution in Ontario, Canada during the COVID-19 state of emergency. Sci Total Environ 742:140516. https://doi.org/10.1016/j.scitotenv.2020.140516

Adger WN, Barnett J, Brown K, Marshall N, O'Brien K (2013) Cultural dimensions of climate change impacts and adaptation. Nat Clim Chang 3:112–117. https://doi.org/10.1038/nclimate1666

Al Huraimel K, Alhosani M, Kunhabdulla S, Stietiya MH (2020) SARS-CoV-2 in the environment: modes of transmission, early detection and potential role of pollutions. Sci Total Environ 744:140946. https://doi.org/10.1016/j.scitotenv.2020.140946

Bakar NA, Rosbi S (2020) Effect of coronavirus disease (COVID-19) to tourism industry. Int J Adv Eng Res Sci 7(4):189–193. https://doi.org/10.22161/ijaers.74.23

Baldasano JM (2020) COVID-19 lockdown effects on air quality by NO_2 in the cities of Barcelona and Madrid (Spain). Sci Total Environ 741:140353. https://doi.org/10.1016/j.scitotenv.2020.140353

Barceló D (2020) An environmental and health perspective for COVID-19 outbreak: Meteorology and air quality influence, sewage epidemiology indicator, hospitals disinfection, drug therapies and recommendations. J Environ Chem Eng 8(4):104006. https://doi.org/10.1016/j.jece.2020.104006

Bera B, Bhattacharjee S, Shit PK, Sengupta N, Saha S (2021) Significant impacts of COVID-19 lockdown on urban air pollution in Kolkata (India) and amelioration of environmental health. Environ Dev Sustain 23: 6913–6940. https://doi.org/10.1007/s10668-020-00898-5

Bhowmick GD, Dhar D, Nath D, Ghangrekar MM, Banerjee R, Das S, Chatterjee J (2020) Coronavirus disease 2019 (COVID-19) outbreak: some serious consequences with urban and rural water cycle. Nat Partner J Clean Water 3:32. https://doi.org/10.1038/s41545-020-0079-1

Casado-Aranda LA, Sánchez-Fernández J, Viedma-Del-Jesús MI (2021) Analysis of the scientific production of the effect of COVID-19 on the environment: a bibliometric study. Environ Res 193:110416. https://doi.org/10.1016/j.envres.2020.110416

Chakraborty B, Roy S, Bera A, Adhikary PP, Bera B, Sengupta D, Bhunia GS, Shit PK (2021) Cleaning the river Damodar (India): impact of COVID-19 lockdown on water quality and future rejuvenation strategies. Environ Dev Sustain 23:11975–11989. https://doi.org/10.1007/s10668-020-01152-8

Chen C, Wang J, Kwong J, Kim J, van Donkelaar A, Martin RV, Hystad P, Su Y, Lavigne E, Kirby-McGregor M, Kaufman JS, Benmarhnia T, Chen H (2020) Association between long-term exposure to ambient air pollution and COVID-19 severity: a prospective cohort study. Can Med Assoc J 194(20):E693–E700. https://doi.org/10.1503/cmaj.220068

Cheval S, Adamescu CM, Georgiadis T, Herrnegger M, Piticar A, Legates DR (2020) Observed and potential impacts of the COVID-19 pandemic on the environment. Int J Environ Res Public Health 17(11):4140. https://doi.org/10.3390/ijerph17114140

Chowdhury AJK, John A, Abdullah R, Musa N, Ashraf MA (2021) Impact of COVID-19 on aquaculture and fisheries in ASEAN countries: some aspects of challenges, mitigations for future strategies in Malaysia. Desalin Water Treat 241:331–339

Corlett RT, Primack RB, Devictor V, Maas B, Goswami VR, Bates AE, Koh LP, Regan TJ, Loyola R, Pakeman RJ, Cumming GS, Pidgeon A, Johns D, Roth R (2020) Impacts of the coronavirus pandemic on biodiversity conservation. Biol Cons 246:108571. https://doi.org/10.1016/j.biocon.2020.108571

Dantas G, Siciliano B, França BB, da Silva CM, Arbilla G (2020) The impact of COVID-19 partial lockdown on the air quality of the city of Rio de Janeiro, Brazil. Sci Total Environ 729:139085. https://doi.org/10.1016/j.scitotenv.2020.139085

Dutta V, Dubey D, Kumar S (2020) Cleaning the River Ganga: impact of lockdown on water quality and future implications on river rejuvenation strategies. Sci Total Environ 743:140756. https://doi.org/10.1016/j.scitotenv.2020.140756

Espejo W, Celis JE, Chiang G, Bahamonde P (2020) Environment and COVID-19: pollutants, impacts, dissemination, management and recommendations for facing future epidemic threats. Sci Total Environ 747:141314. https://doi.org/10.1016/j.scitotenv.2020.141314

FAO (2020a) Keeping food and agricultural systems alive: analyses and solutions in response to COVID-19. Accra. https://doi.org/10.4060/ca8822en

FAO (2020b) The effect of COVID-19 on fisheries and aquaculture in Asia. Bangkok. https://doi.org/10.4060/ca9545en

Fattorini D, Regoli F (2020) Role of the chronic air pollution levels in the Covid-19 outbreak risk in Italy. Environ Pollut 264:114732. https://doi.org/10.1016/j.envpol.2020.114732

Ghahremanloo M, Lops Y, Choi Y, Mousavinezhad S (2021) Impact of the COVID-19 outbreak on air pollution levels in East Asia. Sci Total Environ 754:142226. https://doi.org/10.1016/j.scitotenv.2020.142226

Hashim BM, Al-Naseri SK, Al-Maliki A, Al-Ansari N (2021) Impact of COVID-19 lockdown on NO_2, O_3, PM2.5 and PM10 concentrations and assessing air quality changes in Baghdad, Iraq. Sci Total Environ 754:141978. https://doi.org/10.1016/j.scitotenv.2020.141978

Higham JE, Ramírez CA, Green MA, Morse AP (2021) UK COVID-19 lockdown: 100 days of air pollution reduction? Air Qual Atmos Health 14:325–332. https://doi.org/10.1007/s11869-020-00937-0

Hockings M, Dudley N, Elliott W, Ferreira MN, MacKinnon K, Pasha MKS, Phillips A, Stolton S, Woodley S, Appleton M, Chassot O, Fitzsimons J, Galliers C, Kroner GR, Goodrich J, Hopkins J, Jackson W, Jonas H, Long B, Mumba M, Parrish J, Paxton M, Phua C, Plowright R, Rao M, Redford K, Robinson J, Rodríguez CM, Sandwith T, Spenceley A, Stevens C, Tabor G, Troëng S, Willmore S, Yang A (2020) Covid-19 and protected and conserved areas. Parks 26(1):7–24. https://doi.org/10.2305/IUCN.CH.2020.PARKS-26-1MH.en

I4CE (2020) Investing in climate can help France drive its economic recovery. https://www.i4ce.org/download/investing-in-climate-can-help-france-drive-its-economic-recovery/

Johns Hopkins University Coronavirus Resource Center (2021) COVID-19 dashboard by The Center for Systems Science and Engineering (CSSE) at Johns Hopkins University (JHU). https://coronavirus.jhu.edu/map.html

Ju MJ, Oh J, Choi YH (2021) Changes in air pollution levels after COVID-19 outbreak in Korea. Sci Total Environ 750:141521. https://doi.org/10.1016/j.scitotenv.2020.141521

Kanniah KD, Kamarul Zaman NAF, Kaskaoutis DG, Latif MT (2020) COVID-19's impact on the atmospheric environment in the Southeast Asia region. Sci Total Environ 736:139658. https://doi.org/10.1016/j.scitotenv.2020.139658

Karunanidhi D, Aravinthasamy P, Subramani T, Setia R (2021) Effects of COVID-19 pandemic lockdown on microbial and metals contaminations in a part of Thirumanimuthar River, South India: A comparative health hazard perspective. J Hazard Mater 416:125909. https://doi.org/10.1016/j.jhazmat.2021.125909

Khan R, Saxena A, Shukla S, Selvam S, Goel P (2021) Effect of COVID-19 lockdown on the water quality index of River Gomti, India, with potential hazard of faecal-oral transmission. Environ Sci Pollut Res 28: 33021–33029. https://doi.org/10.1007/s11356-021-13096-1

Kundu C (2020) Has the Covid-19 lockdown returned dolphins and swans to Italian waterways? https://www.indiatoday.in/fact-check/story/has-covid19-lockdown-returned-dolphins-swans-italian-waterways (1658457-2020-03-22)

KWR (2020) What we learn about the Corona virus through waste water research. https://www.kwrwater.nl/en/actueel/what-can-we-learn-about-the-corona-virus-through-waste-water-research

Lal R, Brevik EC, Dawson L, Field D, Glaser B, Hartemink AE, Hatano R, Lascelles B, Monger C, Scholten T, Singh BR, Spiegel H, Terribile F, Basile A, Zhang Y, Horn R, Kosaki T, Sánchez LBR (2020) Managing soils for recovering from the COVID-19 pandemic. Soil Syst 4(3):46. https://doi.org/10.3390/soilsystems4030046

Lokhandwala S, Gautam P (2020) Indirect impact of COVID-19 on environment: a brief study in Indian context. Environ Res 188:109807. https://doi.org/10.1016/j.envres.2020.109807

Mahato S, Pal S, Ghosh KG (2020) Effect of lockdown amid COVID-19 pandemic on air quality of the megacity Delhi, India. Sci Total Environ 730:139086. https://doi.org/10.1016/j.scitotenv.2020.139086

MalayMail (2020) After initial sightings, otters now present at four Klang Valley spots. https://www.malaymail.com/news/malaysia/2020/02/09/after-initial-sightings-otters-now-present-at-four-klang-valley-spots/1835764

Malaysia News Yahoo (2020) More animal crossings: family of elephants spotted near Terengganu petrol station amid MCO. Arif Zikri. https://malaysia.news.yahoo.com/more-animal-crossings-family-elephants-060411933.html

Manenti R, Mori E, Di Canio V, Mercurio S, Picone M, Caffi M, Brambilla M, Ficetola GF, Rubolini D (2020) The good, the bad and the ugly of COVID-19 lock-down effects on wildlife conservation: Insights from the first European locked down country. Biol Cons 249:108728. https://doi.org/10.1016/j.biocon.2020.108728

Manoiu VM, Kubiak-Wójcicka K, Craciun AI, Akman Ç, Akman E (2022) Water quality and water pollution in time of COVID-19: positive and negative repercussions. Water 14:1124. https://doi.org/10.3390/w14071124

Mostafa MK, Gamal G, Wafiq A (2021) The impact of COVID 19 on air pollution levels and other environmental indicators - a case study of Egypt. J Environ Manage 277:111496. https://doi.org/10.1016/j.jenvman.2020.111496

Mousazadeh M, Paital B, Naghdali Z, Mortezania Z, Hashemi M, Niaragh KE, Aghababaei M, Ghorbankhani M, Lichtfouse E, Sillanpää M, Hashim KS, Emamjomeh MM (2021) Positive environmental effects of the coronavirus 2020 episode: a review. Environ Dev Sustain 23:12738–12760. https://doi.org/10.1007/s10668-021-01240-3

Mukherjee P, Pramanick P, Zaman S, Mitra A (2020) Eco-restoration of River Ganga water quality during COVID19 lockdown period using Total Coliform (TC) as a proxy. NUJS J Regul Stud (Special Issue, April), 75–82. https://www.nujs.edu/wp-content/uploads/2022/12/File-85.pdf

Nakada LYK, Urban RC (2020) COVID-19 pandemic: impacts on the air quality during the partial lockdown in São Paulo state, Brazil. Sci Total Environ 730:139087. https://doi.org/10.1016/j.scitotenv.2020.139087

Natural Capital Coalition (2016) Natural capital protocol. https://naturalcapitalcoalition.org/wp-content/uploads/2018/05/NCC_Protocol_WEB_2016-07-12-1.pdf

NRC, National Research Council (US) (1989) Committee on risk perception and communication. National Academies Press (US), Washington (DC)

OECD (2019) Biodiversity: finance and the economic and business case for action—report prepared for the G7 Environment Ministers' meeting. https://www.oecd.org/environment/resources/biodiversity/biodiversity-finance-and-the-economic-and-business-case-for-action.html

OECD (2020) Building back better: a sustainable, resilient recovery after COVID-19. https://www.oecd.org/coronavirus/policy-responses/building-back-better-a-sustainable-resilient-recovery-after-covid-19-52b869f5/

Othman M, Latif MT (2021) Air pollution impacts from COVID-19 pandemic control strategies in Malaysia. J Clean Prod 291:125992. https://doi.org/10.1016/j.jclepro.2021.125992

Pal N, Barman P, Das S, Zaman S, Mitra A (2020) Status of brackish water phytoplankton during COVID-19 lockdown phase. NUJS J Regul Stud 83–86. Special Issue (April). https://www.nujs.edu/wp-content/uploads/2022/12/File-86.pdf

Parrino EL, Falaschi M, Manenti R, Ficetola GF (2021) Lockdown policy effects on invasive species: a perspective. Biodiversity 22(1–2):35–40. https://doi.org/10.1080/14888386.2021.1919554

Pei L, Wang X, Guo B, Guo H, Yu Y (2021) Do air pollutants as well as meteorological factors impact Corona Virus Disease 2019 (COVID-19)? Evidence from China based on the geographical perspective. Environ Sci Pollut Res Int 28(27): 35584–35596. https://doi.org/10.1007/s11356-021-12934-6

Poch RM, Dos Anjos LHC, Attia R, Balks M, Benavides-Mendoza A, Bolaños-Benavides MM, Calzolari C, Chabala LM, de Ruiter PC, Francke-Campaña S, Préchac GF, Graber RE, Halavatau S, Hassan MK, Hien E, Jin K, Khan M, Konyushkova M, Lobb AD, Moshia EM, Murase J, Nziguheba G, Patra KA, Pierzynski G, Eugenio RN, Rojas VR (2020) Soil: the great connector of our lives now and beyond COVID-19. Soil 6(2):541–547. https://doi.org/10.5194/soil-6-541-2020

Praveena SM, Aris AZ (2021) The impacts of COVID-19 on the environmental sustainability: a perspective from the Southeast Asian region. Environ Sci Pollut Res 28:63829–63836. https://doi.org/10.1007/s11356-020-11774-0

Querol X, Massagué J, Alastuey A, Moreno T, Gangoiti G, Mantilla E, Duéguez JJ, Escudero M, Monfort E, García-Pando CP, Petetin H, Jorba O, Vázquez V, de la Rosa J, Campos A, Muñóz M, Monge S, Hervás M, Javato R, Cornide MJ (2021) Lessons from the COVID-19 air pollution decrease in Spain: Now what? Sci Total Environ 779:146380. https://doi.org/10.1016/j.scitotenv.2021.146380

Rawal V, Kumar M, Verma A, Pais J (2020) COVID-19 lockdown: impact on agriculture and rural economy. Social Sci 48:67–82. https://www.jstor.org/stable/26979098

Rume T, Islam SMD (2020) Environmental effects of COVID-19 pandemic and potential strategies of sustainability. Heliyon 6:e04965. https://doi.org/10.1016/j.heliyon.2020.e04965

Rutz C, Loretto M-C, Bates AE, Davidson SC, Duarte CM, Jetz W, Johnson M, Kato A, Kays R, Mueller T, Primack RB, Ropert-Coudert Y, Tucker MA, Wikelski M, Cagnacci F (2020) COVID-19 lockdown allows researchers to quantify the effects of human activity on wildlife. Nat Ecol Evol 4:1156–1159. https://doi.org/10.1038/s41559-020-1237-z

Sax DF, Gaines SD (2008) Species invasions and extinction: the future of native biodiversity on islands. Proc Natl Acad Sci 105(1):11490–11497. https://doi.org/10.1073/pnas.0802290105

Sbai SE, Mejjad N, Norelyaqine A, Bentayeb F (2021) Air quality change during the COVID-19 pandemic lockdown over the Auvergne-Rhône-Alpes region, France. Air Qual Atmos Health 14: 617–628. https://doi.org/10.1007/s11869-020-00965-w

Selvam S, Muthukumar P, Venkatramanan S, Roy PD, Bharath KM, Jesuraja K (2020a) SARS-CoV-2 pandemic lockdown: effects on air quality in the industrialized Gujarat state of India. Sci Total Environ 737:140391. https://doi.org/10.1016/j.scitotenv.2020.140391

Selvam S, Jesuraja K, Venkatramanan S, Chung SY, Roy PD, Muthukumar P, Kumar M (2020b) Imprints of pandemic lockdown on subsurface water quality in the coastal industrial city of Tuticorin, South India: a revival perspective. Sci Total Environ 738:139848. https://doi.org/10.1016/j.scitotenv.2020.139848

Sicard P, De Marco A, Agathokleous E, Feng Z, Xu X, Paoletti E, Rodriguez JJD, Calatayud V (2020) Amplified ozone pollution in cities during the COVID-19 lockdown. Sci Total Environ 735:139542. https://doi.org/10.1016/j.scitotenv.2020.139542

Silva-Rodríguez EA, Gálvez N, Swan GJF, Cusack JJ, Moreira-Arce D (2021) Urban wildlife in times of COVID-19: What can we infer from novel carnivore records in urban areas? Sci Total Environ 765:142713. https://doi.org/10.1016/j.scitotenv.2020.142713

Stratoulias D, Nuthammachot N (2020) Air quality development during the COVID-19 pandemic over a medium-sized urban area in Thailand. Sci Total Environ 746:141320. https://doi.org/10.1016/j.scitotenv.2020.141320

Wang Q, Su M (2020) A preliminary assessment of the impact of COVID-19 on environment—a case study of China. Sci Total Environ 728:138915. https://doi.org/10.1016/j.scitotenv.2020.138915

Wang Y, Yuan Y, Wang Q, Liu C, Zhi Q, Cao J (2020) Changes in air quality related to the control of coronavirus in China: implications for traffic and industrial emissions. Sci Total Environ 731:139133. https://doi.org/10.1016/j.scitotenv.2020.139133

WHO (2020) Novel coronavirus (2019-nCoV): situation report 3. World Health Organization, Geneva, Switzerland, p 2020. https://www.who.int/docs/default-source/Coronaviruse/situation-reports/20200123-sitrep-3-2019-ncov.pdf

WHO (2022) Ambient (outdoor) air pollution. https://www.who.int/news-room/fact-sheets/detail/ambient-(outdoor)-air-quality-and-health

Yao Y, Pan J, Wang W, Liu Z, Kan H, Qiu Y, Meng X, Wang W (2020) Association of particulate matter pollution and case fatality rate of COVID-19 in 49 Chinese cities. Sci Total Environ 741:140396. https://doi.org/10.1016/j.scitotenv.2020.140396

Yunus AP, Masago Y, Hijioka Y (2020)COVID-19 and surface water quality: improved lake water quality during the lockdown. Sci Total Environ 731:139012. https://doi.org/10.1016/j.scitotenv.2020.139012

Yusoff FM, Abdullah AF, Aris AZ, Umi WAD (2021) Impacts of COVID-19 on the aquatic environment and implications on aquatic food production. Sustainability 13:11281. https://doi.org/10.3390/su132011281

Zambrano-Monserrate MA, Ruano MA, Sanchez-Alcalde L (2020) Indirect effects of COVID-19 on the environment. Sci Total Environ 728:138813. https://doi.org/10.1016/j.scitotenv.2020.138813

Zangari S, Hill DT, Charette AT, Mirowsky JE (2020) Air quality changes in New York City during the COVID-19 pandemic. Sci Total Environ 742:140496. https://doi.org/10.1016/j.scitotenv.2020.140496

Zuo M (2020). Coronavirus leaves China with mountains of medical waste, South China Morning Post. https://www.scmp.com/news/china/society/article/3074722/coronavirus-leaves-china-mountains-medical-waste

Chapter 2
Air Quality Changes in Guilin During the COVID-19 Prevention and Control Period

Qing Liu, Juan Wang, and Jiuju Zhang

Abstract The air quality changes and air pollution characteristics in Guilin during the COVID-19 prevention and control period were analyzed by the paired t-test, Wilcoxon signed rank-sum test, chi-square test and pollutant characteristic radar chart using the mean daily AQI and the concentration of 6 air pollutants (PM2.5, PM10, SO_2, NO_2, CO, O_3) from January to March of 2017–2021. The results showed that the air quality in Guilin was generally good. Compared with 2017–2019, the mean daily AQI in 2020 decreased by 29.59%, and the coal burning, dust and motor vehicle pollution reduced, indicating that the air quality was improved significantly. In the strict pandemic prevention and control period, the AQI and the days with PM2.5 pollution decreased by 52.14% and 70.59% respectively, and the days with PM10, PM2.5, SO_2, NO_2 all decreased by more than 50%. Later, the coal and motor vehicle pollution showed signs of increase, suggesting that the type of air pollution changed from combined pollution to coal burning pollution. Compared with 2020, the coal burning, dust and motor vehicle pollution in 2021 experienced a retaliatory rebound, and the air quality basically recovered to the pre-pandemic level, suggesting that the type of air pollution changed from coal burning and motor vehicle pollution to coal burning and dust pollution.

Keywords COVID-19 · Air quality · Guilin · Pollutants · Pollution characteristics

Since the outbreak of COVID-19 in early 2020, China has taken the most comprehensive, rigorous and thorough national COVID-19 prevention and control measures (Xinhua News Agency 2022). Quarantine, a main COVID-19 prevention and control

CLC number: X511 Document code: A

Q. Liu · J. Zhang (✉)
Junior College, Nanning Normal University, Nanning, China
e-mail: 15309152@qq.com

J. Wang
School of Culture and Tourism, Jiangsu University of Technology, Changzhou, China

D. Jones and A. J. Khan Chowdhury (eds.), *Environmental Resilience in the Pandemic Years 2020–2021*, Environmental Science and Engineering,
https://doi.org/10.1007/978-3-031-47757-7_2

measure, can effectively block the transmission of virus (Prem et al. 2020), but it has a negative impact on economic and social development as well as air quality (Wang et al. 2020; Ghahremanloo et al. 2021), which provides opportunities and conditions to observe the relationship between human activities and air quality, and to study the change characteristics and influencing factors of air pollutants.

Some studies have found that, compared with 2017–2019, in 2020: the NO_2 concentration in the continental US decreased significantly by 25.5% (Berman and Ebisu 2020); the concentration of pollutants mainly caused by vehicle emissions in Milan decreased significantly, but the O_3 concentration increased significantly (Collivignarelli et al. 2020); the air quality in medium-sized Thai cities improved significantly (Stratoulias and Nuthammachot 2020). Compared with 2019, in 2020: the NO_2 and PM2.5 concentrations in cities across China decreased significantly and the temporal variation and spatial distribution were relatively consistent, but the O_3 concentration increased (Chu et al. 2021); the air quality in Shijiazhuang significantly improved (Cao et al. 2021); the social activity and NO_2 concentration in Jinan decreased significantly in a stepwise manner (Ai and Wu 2021); the SO_2, PM10 and NO_2 concentrations in the Beijing-Tianjin-Hebei region dropped considerably but the O_3 concentration rose by 76.2% (Zhao et al. 2021). Most of the studies have focused on 2020, believing that the pandemic prevention and control measures have reduced human activities, significantly decreased the concentration of some air pollutants, and significantly improved the air quality, but the increase of O_3 concentration deserves attention. However, some studies have also found that the reduced air pollutant emissions still cannot prevent heavy pollution due to the adverse meteorological conditions (Wang et al. 2020).

Guilin is a well-known international tourism city in Guangxi Province, China, and air quality plays an important role in the development of its tourism (Sun and Qian 2020). The purpose of this study is to explore the air quality changes and air pollution characteristics in Guilin during the COVID-19 prevention and control period, and to preliminarily analyze the main air pollutants influencing the air quality, so as to provide a reference for further optimizing the air quality in Guilin. Compared with previous studies, this study increases the time span to cover the process of economic and social development in Guilin from normal to slowing down to recovery during the COVID-19 prevention and control period, which improves the theoretical reference value of air quality research of Guilin and tourism cities.

2.1 Materials and Methods

2.1.1 Data Source

The mean daily AQI and the concentration of 6 air pollutants (PM2.5, PM10, SO_2, NO_2, CO, O_3) at 6 monitoring sites in Guilin were from https://www.zq12369.com and https://sthjt.gxzf.gov.cn. We adopted the maximum 8 h daily concentration of O_3 as its mean daily concentration, and the mean 24 h daily concentration of other pollutants as their mean daily concentration (Ministry of Environmental Protection 2016).

2.1.2 Research Methods

2.1.2.1 Division of Periods and Stages

On January 24, 2020, Guangxi launched a level I emergency response to combat the outbreak of COVID-19, and all non-necessary transportation and personnel flow were strictly controlled; On February 17, 2020, Guangxi began to take multiple measures to promote the resumption of work and production with precise pandemic prevention and control strategies; Since March 9, 2020, Guilin has been downgraded to a low-risk area, and the resumption of work and production has been launched in full swing, marking the normalized pandemic prevention and control.

According to the study time span and the above time nodes, the pandemic prevention and control period was divided into the outbreak pandemic prevention and control period (2020) and the regular pandemic prevention and control period (2021). The outbreak pandemic prevention and control period was divided into the early pandemic prevention and control stage (stage 1, January 1-January 23), the strict pandemic prevention and control stage (stage 2, January 24–February 17), the preliminary recovery stage (stage 3, February 18–March 8) and the full recovery stage (stage 4, March 9–March 31). The regular pandemic prevention and control period was divided in the same way.

2.1.2.2 Pair Analysis

To reduce the impact of regional pollutant diffusion and meteorological conditions on the study, the data in different years within a certain time span were paired. Specifically, the air quality data from January to March of 2017–2021 were collected and sorted out, and the mean daily AQI and the mean daily concentration of 6 air pollutants in different periods were paired one by one to analyze the air quality differences and change characteristics. Among them, the mean of the same period was used as the pre-pandemic data (from January to March of 2017–2019) to reduce the random error.

Statistical analysis of the data was performed using SPSS 22.0. Normality was tested using K–W test. For normally distributed measurement data, their centralized and discrete trends were described with $\overline{x} \pm s$ using paired t-test. For non-normally distributed measurement data, their centralized and discrete trends were described with median and upper and lower quartiles using Wilcoxon signed rank-sum test, and the count data were tested using chi-square test (Cao et al. 2021; Yang and Zhang 2001). The differences of air quality in different periods were evaluated by the variance ratio, with the formula: $V_{ij} = (A_j - A_i)/A_i \times 100\%$, where V_{ij} was the variance ratio of AQI and the concentration of air pollutants in the j period compared to the i period, A_i and A_j were the AQI and the concentration of air pollutants in the i and j periods, respectively. The significance level was 0.05.

2.1.2.3 Characteristics Radar Chart of Historical Pollutants

Characteristics radar chart analysis of historical pollutants is a research method where the standardized eigenvalue of pollutants and the standardized eigenvalue and upper (lower) limits of corresponding pollutants in a certain period are calculated by mathematical algorithm and drawn on the radar chart to eliminate the influence of air pollutant concentration changes and the differences in different air pollutant concentrations for an intuitive analysis of air pollution characteristics in a specific time (Yang and Zhang 2001; Duan et al. 2018). Please refer to literature Duan et al. (2018) for relevant calculation formula, pollutant source and air pollution type judgment.

2.2 Results and Analysis

2.2.1 Air Quality Changes Before the Pandemic and During the Pandemic Prevention and Control Period

The air quality in Guilin was generally good, with the mean daily AQI below 100 (see Fig. 2.1 for details). Compared with 2017–2019, the mean daily AQI in 2020 fluctuated basically the same, with an overall downward shift, among which the mean daily AQI in Stage 2 showed a significant overall downward shift (Fig. 2.1a and b). The mean daily AQI in 2021 fluctuated largely, among which the mean daily AQI in Stage 4 increased because of the generation of PM2.5 caused by the relative humidity of atmosphere and the robust economic and social recovery after the Spring Festival (Liu et al. 2021) (Fig. 2.1c).

Compared with 2017–2019, the mean daily AQI in 2020 decreased by 29.59% ($Z = -4.89$, $P < 0.05$), indicating that the air quality improved significantly. The mean daily AQI in 2021 did not change significantly ($P > 0.05$), indicating that the air quality basically recovered to the pre-pandemic level (see Table 2.1 for details). Among them, the mean daily AQI in Stage 2 in 2020 fell the most, reaching 52.14% ($Z = -4.08$, $P < 0.05$), indicating that the economic and social activities in Guilin were constrained and the pollutant emissions were reduced under the strict pandemic prevention and control measures; the mean daily AQI in Stages 3 and 4 in 2020 did not change significantly ($P > 0.05$), indicating that the resumption of work and production reduced the air quality. The significant decrease and increase in the mean daily AQI in Stages 2 and 4 in 2021 offset each other, neutralizing the change in the mean daily AQI. Compared with 2020, the mean daily AQI in 2021 increased by 42.39% ($Z = -4.46$, $P < 0.05$), among which the mean daily AQI in Stages 1, 2 and 4 increased by 52.14% ($t = -4.09$), 43.90% ($Z = -3.11$) and 62.76% ($t = -3.73$) ($P < 0.05$), respectively, indicating that the air quality decreased significantly. The reason may be the impact of meteorological conditions and the lack of vigorous air pollution prevention and control measures during the resumption of work and production.

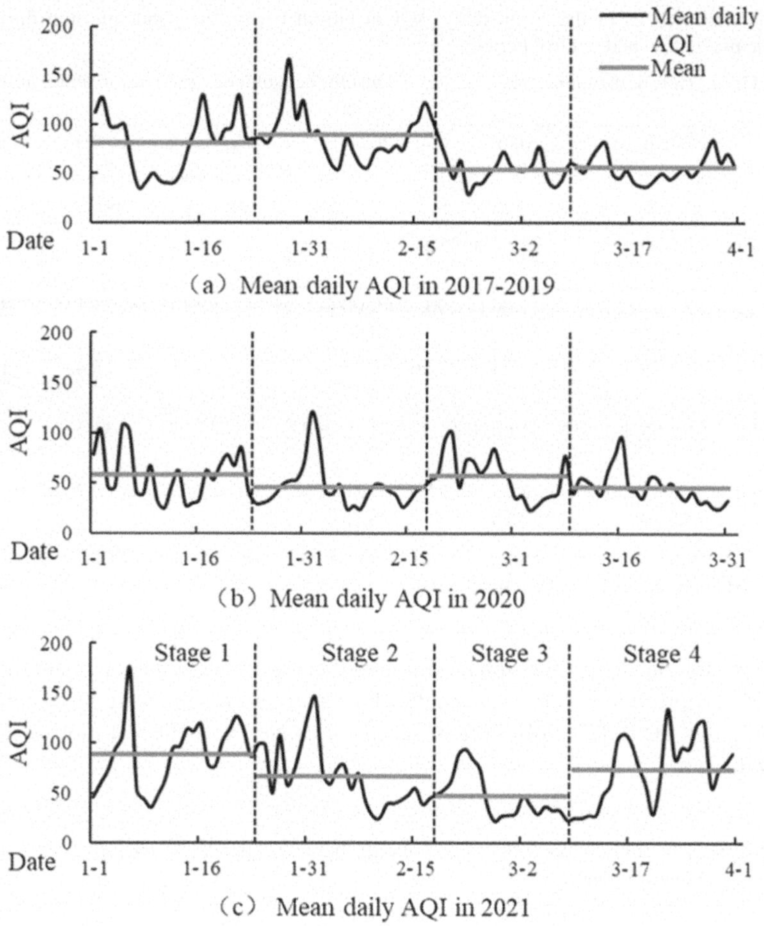

Fig. 2.1 Mean daily AQI in Guilin before the pandemic and during the pandemic prevention and control period

2.2.2 *Comparison of days with Primary Air Pollutant Pollution*

During the study period, the primary air pollutant in Guilin was PM2.5. The number of days with PM2.5 pollution in Stage 2 in 2020 was 5, 12 fewer than 2017–2019, a 70.59% decrease ($\chi^2 = 4.18, P < 0.05$) (see Table 2.2 for details), indicating that both the primary and secondary PM2.5 pollution in Guilin was significantly controlled under the strict pandemic prevention and control measures. The number of days with PM2.5 pollution in 2021 decreased 4 from Stage 3 in 2020 ($\chi^2 = 5.16, P < 0.05$),

Table 2.1 Differences in the mean daily AQI in Guilin before the pandemic and during the pandemic prevention and control period

Stage	Days	Before the pandemic		During the pandemic prevention and control period			
		2017–2019	2020	2021	Variance ratio (%)		
					Variance ratio 1	Variance ratio 2	Variance ratio 3
Stage 1	23	81.10 ± 30.38	58.30 ± 24.86	88.70 ± 33.32	– 28.11*	9.37	52.14*
Stage 2	25	85.67 (72.00, 104.83)	41.00 (31.00, 48.50)	59.00 (42.50, 88.00)	– 52.14*	– 31.13*	43.90*
Stage 3	19	54.23 ± 14.37	56.74 ± 22.56	46.95 ± 23.11	4.63	– 13.42	– 17.25
Stage 4	23	55.65 ± 13.33	45.65 ± 16.42	74.30 ± 33.57	– 17.97	33.51*	62.76*
Total	90	65.33 (50.33, 86.59)	46.00 (34.75, 62.25)	65.50 (41.50, 96.25)	– 29.59*	0.26	42.39*

Note 1: * $P < 0.05$. Note 2: Variance ratios 1, 2 and 3 were the variance of 2020 to 2017–2019, 2021 to 2017–2019, 2021–2020, respectively

and the reason needs further study. Although there were certain changes in the days with PM2.5 pollution in other stages, the chi-square test results were not significant, indicating that the PM2.5 pollution in Guilin reappeared with the resumption of work and production.

Table 2.2 Statistics of the days with PM2.5 pollution before the pandemic and during the pandemic prevention and control period

Stage	Before the pandemic	During the pandemic prevention and control period		Test value					
	2017–2019	2020	2021	2020 and before the pandemic		2021 and before the pandemic		2021 and 2020	
	Days	Days	Days	χ^2	*P*	χ^2	*P*	χ^2	*P*
Stage 1	13	12	18	0.802	0.37	0.449	0.503	0.097	0.756
Stage 2	17	5	14	4.183	**0.041**	1.166	0.28	1.388	0.239
Stage 3	7	10	6	2.949	0.086	0.293	0.588	5.156	**0.023**
Stage 4	8	4	13	0.064	0.801	0.832	0.362	1.859	0.173
Total	45	31	51	6.254	0.100	1.997	0.573	6.751	0.080

2.2.3 Changes in Air Pollutant Concentration

Compared with 2017–2019, the concentration of other 4 air pollutants in 2020 decreased with different degrees except for the insignificant changes in CO and O_3 concentrations ($P > 0.05$) (see Table 2.3 for details). Among them, the SO_2 concentration decreased by 47.04%, and the PM2.5, PM10 and NO_2 concentrations 32.56%, 38.31% and 33.75%, respectively, indicating that under the pandemic prevention and control measures, the coal burning, dust and motor vehicle pollution in Guilin was significantly reduced. This was probably related to the reduction of O_3 precursor such as NO_x (Zhao et al. 2021). The O_3 concentration in 2021 increased by 14.04%, while the concentration of other pollutants did not change significantly, indicating that the emission of air pollutants in Guilin increased with the resumption of work and production. Compared with 2020, the concentration of 6 air pollutants in 2021 all increased ($P < 0.05$), with the top three being SO_2 (91.67%), PM10 (84.93%) and PM2.5 (58.62%), indicating that the coal burning, dust and motor vehicle pollution rebounded significantly.

Compared with 2017–2019, the air pollutant concentration in Stage 2 in 2020 dropped the most ($P < 0.05$) (see Table 2.4 for details), among which the PM10, PM2.5, SO_2, NO_2 and CO concentrations decreased by 63.48% ($t = 7.97$), 60.08% ($t = 7.09$), 56.10% ($t = 12.93$), 53.23% ($t = 7.81$), 14.42% ($t = 2.27$), and the O_3 concentration did not change significantly ($P > 0.05$); the air pollutant concentration in Stages 3 and 4 tended to change insignificantly, with a decreasing decline, indicating that the pollutant emissions tended to increase. Most air pollutant concentrations in 2021 increased, with few decreased ($P < 0.05$). Among them, the O_3

Table 2.3 Changes in the overall air pollutant concentration before the pandemic and during the pandemic prevention and control period

Pollutant	Before the pandemic	During the pandemic prevention and control period		Variance ratio (%)		
	2017–2019 (μg/m^3)	2020 (μg/m^3)	2021 (μg/m^3)	Variance ratio 1	Variance ratio 2	Variance ratio 3
PM2.5	43.00 (32.25, 64.42)	29.00 (16.50, 44.25)	46.00 (23.00, 71.00)	− 32.56*	6.98	58.62*
PM10	59.17 (43.92, 82.75)	36.50 (23.50, 57.25)	67.50 (34.00, 108.00)	− 38.31*	14.08	84.93*
SO_2	11.33 (10.00, 13.67)	6.00 (5.00, 9.25)	11.50 (7.00, 21.00)	− 47.04*	1.50	91.67*
CO#	0.99 (0.83, 1.17)	0.90 (0.78, 1.10)	1.00 (0.90, 1.20)	− 9.09	1.01	11.11*
NO_2	27.17 (23.08, 32.33)	18.00 (13.00, 25.25)	25.00 (16.75, 34.25)	− 33.75*	− 7.99	38.89*
O_3	64.73 ± 16.40	60.96 ± 21.91	73.82 ± 31.43	− 5.82	14.04*	21.10*

Note 1: # was in mg/m^3. * $P < 0.05$. Note 2: Variance ratios 1, 2 and 3 were the variance of 2020 to 2017–2019, 2021 to 2017–2019, 2021 to 2020, respectively

concentration in Stage 1 and the PM10 and SO_2 concentrations in Stage 4 rose the most, with 68.01%, 68.94% and 59.46%, respectively, and the SO_2 concentration in Stage 3 dropped the most, with 32.54%. This may be related to the continued O_3 concentration increase from 2020 to 2021 and the late Spring Festival in 2021. Compared with 2020, the concentration of multiple pollutants in Stages 1, 2 and 4 in 2021 increased significantly ($P < 0.05$), with the top three being SO_2 (110.45%) and PM10 (107.50%) in Stage 2 and PM10 (107.92%) in Stage 4, indicating the great contribution of coal burning and dust pollution to AQI increase in Guilin. The changes in air pollutant concentration during the pandemic prevention and control period initially reflects the main pollution factors affecting the air quality in Guilin.

2.2.4 Characteristics Radar Chart Analysis of Air Pollutants

During the study period, the eigenvalue of the 5 air pollutants in Guilin did not significantly exceed the upper limit, and the upper and lower CO limits were small (see Fig. 2.2 for details), indicating that the type of air pollution was combined pollution (Duan et al. 2018), and the emissions from incomplete combustion of coal burning, motor vehicle fuel and biomass were stable.

The eigenvalue of each pollutant in Stage 1 in 2017–2019 was close to the standard value, and the PM2.5 and PM10 eigenvalues in Stage 2 to Stage 4 in 2017–2019 increased and then gradually decreased to near the standard value, and the NO_2 eigenvalue gradually increased to close to the upper limit. It shows that the air quality in Guilin has the historical characteristics of being mainly affected by fireworks discharge and dust pollution during the Spring Festival and then by motor vehicle pollution after the Spring Festival.

The radar chart in Stage 1 in 2020 was basically the same as that in 2017–2019; The NO_2 eigenvalue in Stage 2 in 2020 was reduced to close to the lower limit, the PM2.5, PM10 and SO_2 eigenvalues were reduced to between the standard value and the lower limit, and the impact of other 4 pollutants except CO was the lowest and the motor vehicle emissions were significantly reduced; The PM2.5 and PM10 eigenvalues from Stage 3 to Stage 4 in 2020 increased significantly to close to the upper limit and then decreased, the change of NO_2 eigenvalue was similar to that in 2017–2019, and the SO_2 eigenvalue increased rapidly to close to the upper limit, suggesting that the type of air pollution changed from combined pollution to coal burning and motor vehicle pollution.

The PM2.5, NO_2 and SO_2 eigenvalues in Stage 1 in 2021 were close to the upper limit, suggesting that the type of air pollution was combined pollution of motor vehicle and coal burning; the eigenvalue of other pollutants from Stage 2 to Stage 3 in 2021 was gradually reduced to close to the lower limit except for the slight reduction in NO_2 eigenvalue, which may be related to the decrease in the intensity of economic and social activities in Guilin during the Spring Festival; The SO_2 and PM10 eigenvalues in Stage 4 in 2021 increased rapidly compared with the previous

Table 2.4 Changes in the air pollutant concentration in different stages before the pandemic and during the pandemic prevention and control period

Stage	Pollutant	Before the pandemic	During the pandemic prevention and control period		Variance ratio (%)		
		2017 ~ 2019 (μg/m³)	2020 (μg/m³)	2021 (μg/m³)	Variance ratio 1	Variance ratio 2	Variance ratio 3
Stage 1	PM2.5	58.68	39.57	64.52	− 32.57*	9.95	63.05*
	PM10	72.49	52.00	94.57	− 28.27*	30.46*	81.87*
	SO_2	16.22	10.52	21.65	− 35.14*	33.48*	105.80*
	CO#	1.18	1.21	1.08	2.54	− 8.47	− 10.74
	NO_2	31.84	28.17	36.17	− 11.53	13.60	28.40
	O_3	54.68	45.22	91.87	− 17.30	68.01*	103.16*
Stage 2	PM2.5	65.33	26.08	45.92	− 60.08*	− 29.71*	76.07*
	PM10	83.24	30.40	63.08	− 63.48*	− 24.22*	107.50*
	SO_2	12.21	5.36	11.28	− 56.10*	− 7.62	110.45*
	CO#	1.04	0.89	1.09	− 14.42*	4.81	22.47*
	NO_2	27.11	12.68	23.40	− 53.23*	− 13.68	84.54*
	O_3	68.69	61.76	79.04	− 10.09	15.07*	27.98*
Stage 3	PM2.5	36.14	37.68	27.95	4.26	− 22.66	− 25.82
	PM10	48.28	47.58	39.37	− 1.45	− 18.45	− 17.26
	SO_2	9.68	6.11	6.53	− 36.88*	− 32.54*	6.87
	CO#	0.85	0.85	0.99	0.00	16.47*	16.47*
	NO_2	22.63	19.00	22.00	− 16.04	− 2.78	15.79
	O_3	67.05	71.05	73.68	5.97	9.89	3.70
Stage 4	PM2.5	36.49	27.43	47.91	− 24.83*	31.30*	74.66*
	PM10	50.83	41.30	85.87	− 18.75	68.94*	107.92*
	SO_2	11.10	12.22	17.70	10.09	59.46*	44.84*
	CO#	0.92	0.84	1.07	− 8.70	16.30*	27.38*
	NO_2	29.25	21.13	28.87	− 27.76*	− 1.30	36.63*
	O_3	68.57	67.48	50.22	− 1.59	− 26.76*	− 25.58

Note 1: # was in mg/m³. * $P < 0.05$. Note 2: Variance ratios 1, 2 and 3 were the variance of 2020 to 2017–2019, 2021 to 2017–2019, 2021 to 2020, respectively

stage, indicating that the coal burning and dust emissions should be controlled even under the normalized pandemic prevention and control measures.

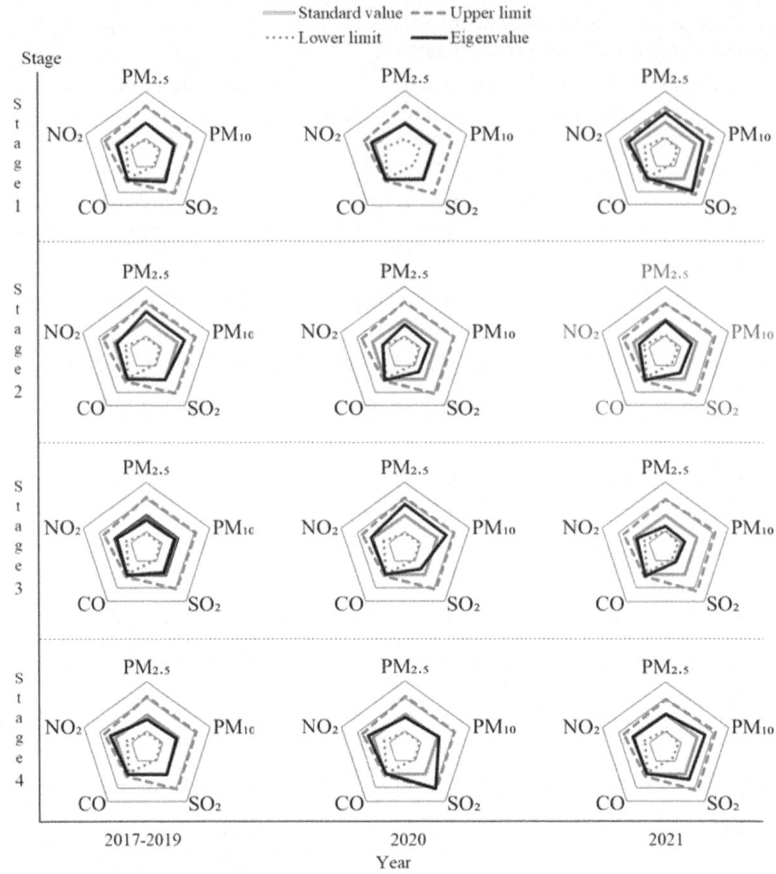

Fig. 2.2 Radar chart of pollutant characteristics before the pandemic and during the pandemic prevention and control period

2.3 Discussion

From 2020 to 2021, the COVID-19 pandemic prevention and control period, the total number of tourists in Guilin increased rapidly, and the mean daily AQI and the concentration of air pollutants also increased rapidly with the fluctuations of enterprise production and transportation. This shows that there is a positive connection between human activities and air quality. In the process of economic and social recovery supported by the tourism in Guilin, the impact of coal burning and dust pollution is significant, followed by the motor vehicle pollution.

Fresh and clean air has always been the basic factor of the development of tourism in Guilin (Sun and Qian 2020). The prevention and control of air pollution is an issue that cannot be ignored in the process of building Guilin into "an international tourist resort". The coal is widely used (Huang et al. 2019), heavy trucks, large buses and medium trucks are main mobile pollution sources (Yin et al. 2017), and the pollutants in the surrounding heavy industrial cities diffuse (Ye et al. 2019), making Guilin still face the air pollution problem. The changes of air quality in Guilin during the pandemic prevention and control period not only show the positive significance of the reduction of human activities to the improvement of air quality, but also show the main artificial source of air pollution in Guilin through the increased and continued impact of pollutants resulted from the centralized resumption of work and production. In the future, it is the main way for Guilin to further optimize the air quality and promote the healthy development of tourism and economic and social development to optimize the energy structure, reduce coal and biomass combustion, vigorously promote new energy vehicles, control construction sites and road dust, and increase the supervision and prevention of air quality.

2.4 Conclusion

(1) Compared with 2017–2019, the coal burning, dust and motor vehicle pollution in 2020 reduced significantly, and the days with PM2.5 pollution decreased, indicating that the air quality was significantly improved. Compared with 2020, the coal burning, dust and motor vehicle pollution in 2021 experienced a retaliatory rebound, and the air quality basically recovered to the pre-pandemic level.

(2) The AQI and the concentration of air pollutants in the strict pandemic prevention and control stage in 2020 significantly decreased, and the number of days with PM2.5 pollution decreased the most, but the coal burning and motor vehicle pollution showed signs of increase in the mid-and post-2020 stages. Compared with 2020, the SO_2, O_3 and PM10 concentrations in 2021 increased by more than 100% in some stages, and the concentration of other pollutants also increased in different stages, indicating that the air pollution rebound in the mid-and post-2020 stage reappeared in the regular pandemic prevention and control period.

(3) The type of air pollution before the pandemic was combined pollution, which changed to coal burning and motor vehicle pollution in 2020, and to coal burning and dust pollution in 2021.

Acknowledgements The https://www.zq12369.com and https://sthjt.gxzf.gov.cn provided the mean daily AQI and the concentration of 6 air pollutants in Guilin, and the website of Guilin Bureau of Statistics provided the economic and social statistics.

Funding Research on the coupling development power of typical mountain tourism system in Guangxi (NO. 2021KY1936); Case Study on the Characteristic Base Construction of the Study Tourism in Mountain Scenic Spots in Guangxi (NO. YXXH2021011).

References

Ai W, Wu Y (2021) Analysis of air pollution and social activity change in jinan during the COVID-19 pandemic. Changjiang Inf Commun 34(4):29–34

Berman JD, Ebisu K (2020) Changes in U.S. air pollution during the COVID-19 pandemic. Sci Total Environ 739:139864

Cao Q, Song J, Zou L et al (2021) Air quality changes in Shijiazhuang during COVID-19 outbreak. J Environ Occup Med 38(5):494–499

Chu BW, Zhang SP, Liu J et al (2021) Significant concurrent decrease in PM (2.5) and NO_2 concentrations in China during COVID-19 epidemic. J Environ Sci 99(1):346–353

Collivignarelli MC, Abba A, Bertanza G et al (2020) Lockdown for CoViD-2019 in Milan: what are the effects on air quality? Sci Total Environ 732:139280

Duan J, Hu J, Tan J et al (2018) Design of characteristic radar chart and its application in air pollution analysis. Res Environ Sci 31(8):1329–1336

Ghahremanloo M, Lops Y, Chol Y et al (2021) Impact of the COVID-19 outbreak on air pollution levels in East Asia. Sci Total Environ 754:142226

Huang J, Chen Z, Mo Z et al (2019) Analysis of characteristics of water-soluble Ions in PM2.5 in Chengdu based on the MARGA. China Environ Sci 39(4):1390–1404

Liu Y, Wang Z, Sun K et al (2021) Analysis of air quality characteristics in Jinan during the COVID-19 prevention and control period. Environ Pollut Cont 43(11):1421–1425

Ministry of Environmental Protection, General Administration of Quality Supervision (2016) Ambient Air Quality Standards: GB 3095-2012. China Environmental Science Press, Beijing

Prem K, Liu Y, Russell TW et al (2020) The effect of control strategies to reduce social mixing on outcomes of the COVID-19 epidemic in Wuhan, China: a modelling study. Lancet Public Health 5(5):261–270

Stratoulias D, Nuthammachot N (2020) Air quality development during the COVID-19 pandemic over a medium-sized urban area in Thailand. Sci Total Environ 746:141320

Sun G, Qian Q (2020) Impact of air pollution on urban tourism development in china—empirical evidence from 168 key monitored cities. J Sichuan Normal Univ 47(5):65–73

Wang PF, Chen KY, Zhu SQ et al (2020) Severe air pollution events not avoided by reduced anthropogenic activities during COVID-19 outbreak. Resource Conserve Recycle 158:104814

Xinhua News Agency (2022) Action in China to fight the COVID-19 epidemic. http://www.xinhuanet.com/2020-06/07/c_1126083364.htm

Yang J, Zhang G (2001) Selection of experimental design scheme and its statistical analysis methods. Chin J Phys Med Rehabil 6:57–60

Ye S, Guan Y, Li S et al (2019) Analysis of aerosol properties in Guilin based on multisource data. Environ Sci Technol 42(10):27–33

Yin Y, Liao L, Luo H (2017) Study of mobile source contribution to the atmospheric fine particulate matter in Guilin city. Environ Protect Sci 43(1):96–99

Zhao X, Shen N, Li L et al (2021) Analysis of changes and factors influencing air pollutants in the Beijing-Tianjin-Hebei region during the COVID-19 prevention and control period. Environ Sci 42(3):1205–1214

Chapter 3
Agricultural-Food Systems and Environmental Health in the Era of Covid-19

Nur Hidayah Hamidi, Osumanu Haruna Ahmed, Adiza Alhassan Musah, Cristalina Jalil Marsal, Syahirah Shahlehi, Khairul Hidayatullah Basir, and Mohd Hamzah Mohd Nasir

Abstract The emergence of novel coronavirus in Wuhan city of China has become a global health emergency. This health pandemic has been affecting normal daily life in every sector around the world particularly the agriculture-food ecosystem. Therefore, this review presents information on the effects of the ongoing global pandemic on the agricultural-food ecosystems and the probable environmental resilience and impact during COVID-19 pandemic years. The COVID-19 pandemic being the most recent in the line of coronavirus zoonotic infections, has created a severe disruption never before in the human history because it affects all work of life including exploiting

N. H. Hamidi
Department of Crop Science, Faculty of Agricultural Science and Forestry, Universiti Putra Malaysia, Bintulu Sarawak Campus, 97008 Bintulu, Malaysia
e-mail: dayahmidi@gmail.com

O. H. Ahmed (✉) · C. J. Marsal · S. Shahlehi · K. H. Basir
Faculty of Agriculture, Universiti Islam Sultan Sharif Ali (UNISSA), KM 33, Jalan Tutong, Kampung Sinaut, Tutong TB1741, Brunei Darussalam
e-mail: ahmed.haruna@unissa.edu.bn

C. J. Marsal
e-mail: cristalina.jalil@unissa.edu.bn

S. Shahlehi
e-mail: syahirah.shahlehi@unissa.edu.bn

K. H. Basir
e-mail: hidayatullah.basir@unissa.edu.bn

A. A. Musah
Graduate School Management, Postgraduate Centre, Management and Science University, University Drive, Off Persiaran Olahraga Section 13, 40100 Shah Alam, Selangor, Malaysia
e-mail: adiza_alhassan@msu.edu.my

M. H. M. Nasir
Department of Biotechnology, Kulliyyah of Science, International Islamic University Malaysia, 25200 Kuantan, Pahang Darul Makmur, Malaysia
e-mail: hamzahn@iium.edu.my

D. Jones and A. J. Khan Chowdhury (eds.), *Environmental Resilience in the Pandemic Years 2020–2021*, Environmental Science and Engineering,
https://doi.org/10.1007/978-3-031-47757-7_3

social, economic, institutional, and environment vulnerabilities and intensify food insecurity, yet pollution is declining and nature is recovering its natural state. Preventive measures implemented by many countries are likely to affect vulnerable groups including workers, small scale farmers, low-income families, and students faced the worst hit among all. Waste generation is in the course of the pandemic is alarming, if not worrying. Educational institutions are searching ways to fill the loss of learning. The COVID-19 pandemic has given us the opportunity to introduce online learning. The COVID-19 pandemic calls for urgent policy solutions to keep the economy afloat and people to keep their jobs and incomes. Although the pandemic creates some major issues for the food system, it also presents a good chance to speed up transformation in the food and agriculture industry that will improve its resilience to a variety of threats including climate change.

Keywords Health pandemic · Supply and demand · Safety · Wastes · Agricultural and food systems · Pollution

3.1 Introduction

In Wuhan city of China, an acute respiratory infection of unknown origin was first reported in December 2019 (Singhal 2020) after which it was termed novel coronavirus infected pneumonia (NICP) (Wang et al. 2020). Through unbiased sequencing, the pathogen was later identified as a novel enveloped RNA β-coronavirus with a similar phylogeny to SAR-CoV (Lai et al. 2020; Zahra et al. 2020). In January 2020, the World Health Organization (WHO) named this new virus as the 2019 novel coronavirus (2019-nCoV) and announced the virus as a worldwide pandemic. In February 2020, the World Health Organization (WHO) named the novel disease as Corona Virus Disease 2019 (COVID-19). Gorbalenya et al. (2020) reported that the Coronavirus Study Group (CSG) named this virus as severe acute respiratory syndrome coronavirus 2 (SARS-CoV-2). According to Chitrakar et al. (2021) the major route of transmission of the virus is person-to-person contact caused by microscopic droplets from an infected person's breathing, sneezing, or coughing. It is believed that these droplets that land on food surfaces, food packaging, or items such as doorknobs and elevator buttons stay there for a while depending on the temperature and make-up of the surfaces. These viruses can spread to people by touching these surfaces and then the nose, eyes, or mouth. In 2020, the SARS-CoV-2 spread rapidly across the world affecting health, food, agriculture, economics, and livelihood (di Marco et al. 2020; Laborde et al. 2020). The COVID-19 is the second pandemic of the twenty-first century after the Middle East Respiratory Syndrome (MERS) caused the largest quarantine in human history (Zowalaty and Järhult 2020). Strong public health measures were taken in response to COVID-19's high level of infectiousness, drawing in part from the lessons learned from previous viral pandemics. These measures included restrictions on local, domestic, and international travel as well as the promotion of good hygiene and social seclusion (Peeri et al. 2020; CCSA

2020). The COVID-19 quarantine imposition owing to the closure of all marketplaces, places of worship, places of public assembly, travels, construction activities, and the global economy and stock market, is unprecedented in the human history. Nonetheless, at the end of 2020, severe outbreaks were still occurring throughout Asia, and the infections were still spreading over the world. The pandemic occurred at a period when numerous issues with sustainable development were becoming more serious (FAO 2020a; Rockstrom et al. 2020; Otsuka and Fan 2021). It was in the light of these challenges that policymakers anticipated that COVID-19 would significantly reduce productivity and food security, particularly for poor rural populations (HLPE 2020a; UNESCAP 2020). Early estimates indicated that the pandemic could cause a major contractions in the world and many national economies (World Bank 2020) as well as an increase in poverty and a doubling of the population that is critically malnourished (FAO 2020a; FSIN 2020; HLPE 2020b). In addition to the implementation of mobility restrictions and community lockdowns during quarantines led to disturbances in the agricultural food systems, which led to supply and demand shocks. This would have an effect on the immediate and long-term economic performance of the agricultural production systems and their contribution to food security. Nevertheless, the COVID-19 pandemic has many good impacts on the climate, which the world experienced while in quarantine. A substantial change in the environment as a result of the quarantine measures put in place by governments worldwide is a good indication for the fatal global environmental catastrophes, such as greenhouse gas emissions and ozone layer depletion. Therefore, this review presents information on the effects of the ongoing global pandemic on the agricultural-food ecosystems and the probable environmental resilience and impact during COVID-19 pandemic years.

3.2 Effects of the Pandemic on the Food Supply Chain and Food Safety

The COVID-19 preventative methods such as mass quarantine, stay-at-home order, lockdown, and transportation halt were incredibly challenging. Even before the COVID-19 crisis, approximately 113 million people worldwide suffered from chronic food insecurity as a consequence of previous shocks (World Bank 2020). This suggests that the several countries in the world were already on the verge of starvation, and were unable to combat the virus. Currently many countries are dealing with several challenges such as natural disasters including flood and drought events, crop destruction, extreme weather, political turmoil, which worsen the situation with regard to food insecurity (Phillipson et al. 2020). The effects will be particularly severe in conflict-affected countries such as Syria, where transportation and supply are already difficult even in the absence of disease and social isolation. When the first case was confirmed, many developing countries declared a state of emergency.

Supply and demand shocks brought on by the COVID-19 pandemic have an impact on all major economic sectors, most notably being on agriculture (Gregorioa and Ancog 2020a, b). Imports and exports moved slower or, in some circumstances, not at all. During this economic recession, countries counted on emergency aid, which perhaps were sufficient to safeguard them and their vulnerable populations. The transport sector almost ceased operations as a result of the lockdowns imposed in some countries. The consequences being disrupting the supply chain for vital or essential items, including food (Reardon et al. 2020) and humanitarian relief supplied by various organizations. Temporary or seasonal employment is widespread in developing and undeveloped nations, particularly for planting, sorting, harvesting, processing, and transporting products to markets. Therefore, the absence of local or migratory workers because of illness or travel restrictions imposed by multiple lockdowns had tremendous impact on the supply chain. In circumstances where the illness directly affects their health or mobility, it also reduced not only the capacity for others to produce but also their own food safety (FAO 2020b). Particularly, the COVID-19 crisis's worker shortage led to serious disruptions in various worker-intensive agricultural industries such as cattle farming, horticulture, planting, harvesting, and agricultural processing (Stephens et al. 2020a, b). It must be stressed that there was a severe problem with the lack of farmworkers even before the COVID-19 outbreak (Richards and Rickard 2020). So the COVID-19 outbreak "was the last straw that broke the camel's back".

The whole food system, the from primary supply to the final demand is disrupted by the COVID-19 pandemic. The scope of COVID-19's effects might be better comprehended across the value chain, which depicts how food flows from farmer-producers to processors, distributors, retailers, and lastly to customers' plates. Generally, the food supply chain demonstrates the relationship between supply and demand in an agricultural food system (Gregorioa and Ancog 2020a, b). The impact on agriculture and food demand includes a variety of macroeconomic factors, substantial volatility in credit markets, exchange rates, energy, and, most importantly, the predicted rise in unemployment and contraction in total economic activity. Furthermore, a breakdown situation in supply chains among many interconnected parties (workers, enterprises, suppliers, customers, and financial intermediaries) is envisaged, with a cascading effect on the market economy. The direct implications of COVID-19 could include reduced employment availability for agricultural operations and regulations to stop the virus from spreading within communities, safeguard susceptible individuals, and promote agriculture (Mandal et al. 2020; Stephens et al. 2020a, b). According to Amjath-Babu et al. (2020), higher disposable income for agricultural households because of welfare programmes, disruption of markets brought on by movement limitations, and policy and programme assistance for agricultural production and selling were all anticipated as indirect consequences of COVID-19.

Despite medical care, food and groceries are permitted considering that they are essentials during quarantine periods. However, the fear of the virus in addition to the restrictions on transportation had a significant impact on food-related operations such as pre-harvest agriculture, post-harvest processing, and distribution. Food was stocked in response to the fear of a scarcity and this led to a temporary market shortage

(Chitrakar et al. 2021). Food safety is important because according to Galanakis (2020), it is one of the four main components of the food system that has been impacted by the coronavirus pandemic. During the pandemic, food safety measures were implemented with an emphasis on the most vital precautions required at each stage of the food supply chain that is, from farm to plate. The food safety measures include personal hygiene, surface disinfection, clean working area, food preparation and delivery, and social distancing. Therefore in this era of the global pandemic, it is important for the food industry to ensure that the food reaching consumers' plates is safe and does not jeopardize their health at any stage of the process, including delivery. This is because fresh food products such as fruits, vegetables or food packaging could be contaminated with the coronavirus if an infected person sneezes or coughs on them. It appears that transmission is conceivable if the virus is transmitted to the mucous membranes of the mouth, throat, or eyes through hands or the meal itself. These implications pose risks to the workers in the food processing industries, potentially disrupting their operations.

The restriction on outdoor activities, along with transportation and travel restrictions, generates a worker and input supply constraint, affecting activities in the food supply chain such as harvesting, processing, and distributing, with the prospect of a food crisis. Ready-to-harvest fresh products are unable to reach market destinations hence, resulting in rotting on fields and a temporary short-supply and long-term food crisis (Chitrakar et al. 2021). Additionally, Galanakis (2020) reported that the consequences of the COVID-19 pandemic with food safety, economy, and humanity are still unravelled. Furthermore, the novel coronavirus is new to the world and scientific research into it is still at its infancy stages. Literature has it that earlier outbreaks of coronaviruses like MERS and SARS-CoV did not spread through food because of the stomach's acidic environment (pH 3.5) (Aday and Aday 2020). However, some cooking and eating practices may cause the coronavirus to emerge in humans from animals (Rizou et al. 2020). Although it was previously stated that food was unlikely to constitute a medium for virus transmission of COVID-19 pandemic marked the beginning of a new era in the food supply chain and industry because researchers and professionals in the food industry encounter challenges including assuring food safety, identifying SARS-CoV-2 in places where food is produced, processed, and distributed, effectively sanitizing surfaces and working spaces. Thus, more precautions are required as the supply chain progresses to the final stages as more people are involved in the process. The risk of COVID-19 transmission through the food business is currently considered low, but governmental agencies do not consider tracing SARS-CoV-2 in the food sector and associated ecosystems to be a priority (Rizou et al. 2020).

3.3 Effects of the Pandemic on Agricultural Food and Other Solid Wastes

Food waste has gained more attention than ever before the era of pandemic. One-third of all food produced is wasted at various phases of the food supply chain including production, postharvest handling, processing, distribution, and consumption even before the pandemic. In the new era of the coronavirus pandemic, food waste garnered more attention than ever before (Aday and Aday 2020). This could be attributed to the consumer behaviour because the pandemic disrupted the daily routine leading to boredom and restlessness described as consuming significant number of calories from fat, carbohydrate, and proteins. Moreover, the multiple quarantines/lockdowns stressed people up and drove them toward sugary foods to remain feeling good. This is because foods rich in carbohydrates can be utilised as components of self-medication because of their ability to promote serotonin production. However, Muscogiuri et al. (2020) reported that these unhealthy eating practices could lead to obesity which is connected to chronic inflammation and serious COVID-19 complications.

The solid waste is more serious in this period of pandemic because people around the world have been required to consume significant amounts of personal protective equipment (PPE). Among these, the use of face masks, face shields, and gloves has increased as an effective approach to prevent the transmission of the virus (Torres and De-la-Torre 2021). As most of these PPEs contain plastic derivatives and plastic, prolonged use produces a significant amount of waste that is released to the environment. According to Prata et al. (2020), approximately 129 billion face masks are used monthly worldwide, with most of these being disposable face masks made up of plastic microfibers. Additionally, according to a WHO research, 89 million medical masks will be required for the United States alone to combat COVID-19 whereas the plastic innovation hub estimates that domestic demand for masks in the United Kingdom is approximately 24 billion per year (Selvaranjan et al. 2021). Furthermore, the Japanese Ministry of Economy, Trade, and Industry stated that approximately 600 million face masks were needed in April 2020 (Fadare and Okoffo 2020), while in Victoria, Australia it is estimated that daily face mask used were 5 million face masks per day (Boroujeni et al. 2021). Similarly, the world's largest manufacturer of face masks, China has increased its daily production of medical face masks to 15 million since February 2020. In most cases, face masks are discarded after single usage and result in polluting the ecosystems such as soil, rivers, oceans, and street by surface runoff, river flows, wind, animals, and oceanic currents through entanglement or ingestion (Xu and Ren 2021). Disposable face masks may release and develop hazardous chemical and biological contaminants such as heavy metals and bisphenol A, similar to other plastic wastes (Xu and Ren 2021). However, many countries in the world, particularly those in Asia have no specific waste mask collection systems (Selvaranjan et al. 2021). Generally, masks in the trash are treated as a general waste which implies that most of the masks will eventually get into the aquatic system, soil system, or being burned with other medical waste. Bondaroff and Cooke (2020) in a research reported that 1.6 million face masks entered water bodies in 2020 and this in the long term could have an impact on the marine ecosystem (De-le-Torre and Aragaw 2021; Fadare and Okoffo 2020).

3.4 Effects of the Pandemic on Agriculture

Agriculture employs approximately one billion people worldwide and not doubt that it is the backbone of many developing countries. The precautionary measures implemented to combat the pandemic impede agricultural product production and distribution. It is possible that the restrictions imposed affect utility supply and demand, thus most countries have designated agriculture and agro-food as essential industries, exempting them from movement restrictions and factory closure. Therefore, the direct effects of the pandemic on primary agriculture should be minimal in many countries because the coronavirus has minimal effect on the natural resources that support production. However, agricultural production is a lengthy process that includes effort at several phases, such as planting, nurturing, harvesting, and commodities export. As a result of their reliance on market value chains, the food and agricultural sectors are regarded as less resilient. The travel restrictions enforced by countries in the world to combat the spread of coronavirus retarded these operations. Agriculture is a high-input sector which is affected by international trade disruption. Agricultural production systems are experiencing a shortage of agricultural inputs such as grain, fertilizer, and pesticides. For example, China is one of the world's biggest manufacturers and exporters of fertilizer. The lockdown in China had a significant influence on the global fertilizer trade (Marlow 2020). In the agricultural systems, the reliance on each factor of production differ significantly as such, it exposes producers and farmers to surges in input costs. These are the results of several factors such as overcrowded ports or highways, delays in customs clearance, interruptions in transportation, a lack of credit availability, and higher interest rates and capital prices. These disruptions increases the agricultural input costs and the cost of production in general. These additional expenses reduce the profitability of agricultural product production and result in a significant loss for agricultural producers (Workie et al. 2020). The on-going COVID-19 pandemic poses a significant threat to food security and livelihood, particularly in the developing and underdeveloped countries where agricultural production systems are more labour intensive and have less capacity to withstand a major socioeconomic catastrophe.

Gregorioa and Ancog (2020a, b) reported that in Southeast Asian, agriculture continues to be an important source of economy, especially in terms of eradicating poverty and ensuring food security. Overall, the region's agricultural production has been on the ascendance, especially in terms of yield and area. The average proportion of agricultural workers to the entire population remains high although the contribution of the agriculture sector to overall output has generally been declining in several Southeast Asian countries. As of 2019, 31% of people in Southeast Asia (SEA) are employed by the agricultural sector. Except Singapore, Malaysia, and Brunei, at least 23% of the workforce is employed by agricultural sector in the rest Southeast Asian countries. The pandemic movement restrictions imposed by community lockdowns result in a reduction in the amount of agriculture workers and correlates with a reduction in overall agricultural output. Moreover, farmers' limited access to farm inputs and, markets to sell products contribute to reduction in agricultural production, the resultant being financial losses and agriculture food wastes. In the long term, loss of income and economic slowdown affect demand, particularly among farmers. Additionally, workers in the informal sector are vulnerable to the coronavirus infestation because they typically work in congested areas where it is difficult to maintain and practice the safety precautions recommended by the WHO. Besides, staying indoors is not an option because they require money for daily subsistence (Workie et al. 2020).

Nevertheless, the COVID-19 era has significantly increased the demand for organic and local foods because of several restrictions including quarantine and travel restrictions and escalating health concerns. The demand for regional and local agricultural products is greater than that for other goods. People demanded regional nutritious food from local farmers (Rowan and Galanakis 2020). In this new era, the utilization of information and technology in agriculture has changed this industry. Social networking has enabled establishing direct communication between farmers and consumers. Farmers are now aware of their opportunities and effectively use them because farmers were able to communicate with customers and deliver products directly to their houses, hence eliminating the need for agents and middlemen. Now, with direct consumer interaction, farmers get the best price for their efforts. Moreover, in several countries, the small scale farmers joined together to secure fair market pricing, and this strategy has been beneficial to small farmers. The concept of group farming arose as a result of travel constraints. Morsy et al. (2021) reported that client hoarding causes wholesalers to demand a large amount of vegetables when travel restrictions are abolished. Thus small farmers work together to satisfy the demand for vegetables with a fair price. There was a surge in demand for dried vegetables including mushrooms and frozen vegetables. By seizing this opportunity, the farmers were able to prevent loss of vegetables and income. Furthermore, there is a considerable demand for milk and related products. Farmers with livestock had an opportunity during COVID-19 pandemic because of the increased demand for milk products and immune booster food (Farell et al. 2020). Moreover, as people were compelled to stay at home, the idea of an individual housetop farm and in-home nurseries increased. The cultivation practices of planting flowers and vegetables have also gained significant importance (Kalogiannidis and Melfou 2020).

3.5 Effects of Pandemic on Education System

The United Nations Educational, Scientific, and Cultural Organization (UNESCO) has stated that the coronavirus pandemic had influenced on the global education system (UNESCO 2020). Following the global COVID-19 outbreak, the governments of several nations made the decision to close all educational institutions, including schools, colleges, and universities, in an effort to create a social barrier and stop the spread of this infectious disease. Approximately 87% of children's education suffered when educational institutions closed during pandemic of COVID-19, affecting 1.5 billion students in 195 countries (UNESCO 2020). The pandemic generates greater educational disturbance and widens educational gaps. School closures can exacerbate existing inequities among students (Owusu-Fordjour et al. 2015). In developing countries, the school closure brings issues for students, families, and teachers. Children with low learning motivation and students from low-income homes suffer the most from coronavirus. The toddlers may have higher dependence on their parents than teenagers and they need guidance in their learning process, internet usage, and use of digital gadgets and applications (Tzifopoulos 2020). Mustafa (2020) reported that the school closures around the world had a severe impact on the education. It knows no bounds and has an immediate impact. Online learning has been implemented in this new era of pandemic, and increased use of internet technology for collecting educational materials (Fry 2001).

Before the pandemic of COVID-19, educational institutions that had implemented new technologies in their operations had a competitive advantage over those that yet implement the technology. During the pandemic, platforms, educational software and resources included in distance learning solutions consisted of massive open online courses, digital learning management systems, and self-directed learning content are designed to empowering teachers, parents, and students. Students and teachers had to adjust new teaching and learning methods. Students in areas where there were no appropriate technologies and facilities to accommodate online education experienced challenges in making the transition to online learning, particularly for students in rural areas. This is because, normally students and teachers in rural areas often lack of the facilities and necessary expertise to implement the online teaching–learning process. In some developing countries, their educational programs are conducted through radio, television, and online platforms. However, the poor families and students have no access to radio, television, and other devices needed to learn at home. Moreover, people who lack in digital skills needed to incorporate with online learning. Technology is still employed to bridge developmental gaps caused by unexpected school cancellation and closure during pandemic. Students, teachers, and faculty members confront new challenges and difficulties in this global fight to combat the pandemic. There are several reasons including restricted learning resources, specific training required, limited wireless internet, and educational interruption. In addition, school closures also include non-school issues such as housing insecurity, health and physiological emotions and other problems that caused the inability to obtain desired educational results. Abari and Orunbon (2020) revealed that school closures have serious psychological, cultural, and economic impacts, and the changes affect people from all profession, but they have a significantly negative impact on the poor. In most countries, online teaching is new and it is not as easily learned as in the classroom. In an online class, a digital education framework needs a high speed of internet connections and convenient location at home. This is because many students in online classes are not paying attention. This new era of digital learning brought in many technological challenges psychological shifts. In other words neither teachers nor students were well prepared for e-learning, which contributed to COVID-19 impacts on educational system.

During the closures, researchers, curriculum designers, education officers, and educational institutions collaborated to improve the education system. Curriculum design, collaborations, skill development, and educational institutions should focus on advancing the education system. For example, curriculums, learning approaches, and the education system itself should be transformed and developed. Tiruneh (2020) stated that after COVID-19, the school should designed techniques and procedures to make up for lost learning, ensuring that students return to school when the schools reopen, and scale distant learning accessibility.

3.6 Benefits of the Pandemic COVID-19 on Aquatic Life, Air Quality, and Climate

Khan et al. (2021) believe that when humanity retreats indoors, the natural world rumbles freely. This is because of the coronavirus, notoriously polluted streams and rivers become cleaner, the smog and haze dispersed or gone, the atmosphere become fresher, and the wildlife has filled the open spaces. The natural world has continued to rumble on while millions of humans have been locked indoors, and the world is thriving in our absence. The pandemic induced lockdown improved water quality in rivers and water bodies. This could be attributed to the cessation of the discharge of industrial effluents and other wastes into water led to improved water quality. After two months of COVID-19 quarantine, the water in Venice become clearer, and the aquatic species that has not been seen in the cities for many years is now visible (Khan et al. 2021).

Additionally, since the quarantine was implemented, many species are returning to their natural habitats because of the clean rivers and water bodies. The closure of factories and businesses globally with the suspension of numerous cruise ships, the slowdown tourism, and the suspension of all other maritime operations has caused the level of pollution to decrease. GeoNoise (2020) wrote that the aquatic life is disturbed by noise levels from shipping traffic, which typically ranges from 20 to 200 Hz and it is reduced by six dicibels with a major reduction below 150 Hz. This is corroborates with the study done by Rolland et al. (2012) from which they stated that marine creatures' stress hormone levels increase in response to ambient noise from maritime traffic, which may have an impact on their reproductive success.

The WHO (2019) reported that outdoor air pollution kills an estimated seven million people annually worldwide, and approximately 80% of the urban population is exposed to polluted air. One of the most harmful pollutants, PM2.5 is a type of particulate matter (PM), classified as a group 1 carcinogen. The 2.5 indicates the particle size (in microns) which is approximately one-thirtieth of the width of a human hair (Xu and Ren 2019). The PM2.5 can pass-through the lungs to the bloodstream because of its small size. Apart from cardiovascular disease and respiratory issues, the PM2.5 causes early death as well as. According to the WHO (2019), PM2.5 causes approximately four million fatalities worldwide annually, including those due to heart disease, strokes, lung cancer, chronic lung disease, and respiratory infections. Besides, the engines of automobiles and factories release harmful gas nitrogen dioxide (NO_2) which if the concentration is higher than 200 g/m^3 it might result in respiratory tract inflammation and asthma. During the pandemic lockdown and quarantine orders, the NO_2 concentrations in most cities reduced significantly from 5.6 to 0.2 g/m^3 (Otmani et al. 2020).

3.7 Conclusion

The COVID-9 pandemic being the most recent in the line of coronavirus zoonotic infections, has created a severe disruption never before in the human history because it affects all work of life including exploiting social, economic, institutional, and environment vulnerabilities and intensify food insecurity, yet pollution is declining and nature is recovering its natural state. Preventive measures implemented by many countries are likely to affect vulnerable groups including workers, small scale farmers, low-income families, and students faced the worst hit among all. Waste generation is in the course of the pandemic is alarming, if not worrying. Educational institutions are searching ways to fill the loss of learning. The COVID-19 pandemic has given us the opportunity to introduce of online learning. The COVID-19 pandemic calls for urgent policy solutions to keep the economy afloat and people to keep their jobs and incomes. Although the pandemic creates some major issues for the food system, it also presents a good chance to speed up transformation in the food and agriculture industry that will improve its resilience to a variety of threats including climate change.

References

Abari AO, Orunbon NO (2020) Building bridges and walls: education and COVID-19 in Nigeria. Res J Compar Educ 1(1):39–52

Aday S, Aday MS (2020) Impact of COVID-19 on the food supply chain. Food Qual Safety 4(4):167–180

Amjath-Babu TS, Krupnik TJ, Thilsted SH, McDonald AJ (2020) Key indicators for monitoring food system disruptions caused by the COVID-19 pandemic: insights from Bangladesh towards effective response. Food Secur 12:761–768. https://doi.org/10.1007/s12571-020-01083-2

Bondaroff TP, Cooke S (2020) Masks on the Beach: the Impact of COVID-19 on marine plastic pollution. OceansAsia

Boroujeni M, Saberian M, Li J (2021) Environmental impacts of COVID-19 on Victoria, Australia, witnessed two waves of coronavirus. Environ Sci Pollut Control Ser 28(11):14182–14191. https://doi.org/10.1007/s11356-021-12556-y

CCSA (The Committee for the Coordination of Statistical Activities) (2020) How COVID-19 is changing the world: a statistical perspective, vol II. https://unstats.un.org/unsd/ccsa/documents/COVID19-report-ccsa.pdf

Chitrakar B, Zhang M, Bhandari B (2021) Improvement strategies of food supply chain through novel food processing technologies during COVID-19 pandemic. Food Control 125:108010

De-la-Torre GE, Aragaw TA (2021) What we need to know about PPE associated with the COVID-19 pandemic in the marine environment. Mar Pollut Bull 163:111879. https://doi.org/10.1016/j.marpolbul.2020.111879

di Marco M, Baker ML, Daszak P, de Barro P, Eskew EA, Godde CM, Harwood TD, Herrero M, Hoskins AJ, Johnson E, Karesh WB, Machalaba C, Garcia JN, Paini D, Pirzl R, Smith MS, Zambrana-Torrelio C, Ferrier S (2020) Opinion: sustainable development must account for pandemic risk. Proc Natl Acad Sci 117(8):3888–3892. https://doi.org/10.1073/pnas.2001655117

Fadare OO, Okoffo ED (2020) Covid-19 face masks: a potential source of microplastic fibers in the environment. Sci Total Environ 737:140279 https://doi.org/10.1016/j.scitotenv.2020.140279

FAO (Food and Agriculture Organization of the United Nations) (2020a) The state of food security and nutrition in the world 2020: transforming food systems for affordable healthy diets. http://www.fao.org/3/ca9692en/online/ca9692en.html

FAO (Food and Agriculture Organization of the United Nations) (2020b) Responding to the impact of the COVID-19 outbreak on food value chains through efficient logistics [Online]. http://www.fao.org/3/ca8466en/CA8466EN.pdf. Accessed on December. 23, 2022.

Farrell P, Thow AM, Wate JT, Nonga N, Vatucawaqa P, Brewer T, Sharp MK, Farmery A, Trevena H, Reeve E, Eriksson H, Gonzalez I, Mulcahy G, Eurich JG, Andrew NL (2020) COVID-19 and Pacific food system resilience: opportunities to build a robust response. Food Security 12(4):783–791

Fry K (2001) E-learning markets and providers: some issues and prospects. Educ Training

FSIN (Food Security Information Network) (2020) 2020 global report on food crises: joint analysis for better decisions. http://ebrary.ifpri.org/utils/getfile/collection/p15738coll2/id/133693/filename/133904.pdf

Galanakis CM (2020) The food systems in the era of the coronavirus (COVID-19) pandemic crisis. Foods 9(4):523

GeoNoise (2020) Coronavirus lockdown gives animals a rare breakfrom noise pollution. https://www.geonoise.com/coronavirus-lockdown-gives-animals-a-rare-break-from-noise-pollution/. Accessed 26 Dec 2022

Gorbalenya AE, Baker SC, Baric RS et al (2020) The species severeacute respiratory syndrome-related coronavirus: classifying 2019-NCoV and naming It SARS-CoV-2. Nat Microbiol 5:536–544. https://doi.org/10.1038/s41564-020-0695-z

Gregorioa GB, Ancog RC (2020a) Assessing the impact of the COVID-19 pandemic on agricultural production in Southeast Asia: toward transformative change in agricultural food systems. Asian J Agric Dev 17(1362-2020-1097):1–13

Gregorioa GB, Ancog RC (2020) Assessing the impact of the COVID-19 pandemic on agricultural production in Southeast Asia: Toward transformative change in agricultural food systems. Asian J Agric Dev 17(1362-2020-1097):1–13

HLPE (High Level Panel of Experts on Food Security and Nutrition Committee on World Food Security) (2020b) Interim issues paper on the impact of COVID-19 on food security and nutrition (FSN). Committee on World Food Security. http://www.fao.org/fileadmin/templates/cfs/Docs1920/Chair/HLPE_English.pdf

HLPE (High Level Panel of Experts on Food Security and Nutrition Committee on World Food Security) (2020a) Food security and nutrition: building a global narrative towards 2030. In: HLPE Report 15. committee on world food security. http://www.fao.org/3/ca9731en/ca9731en.pdf

Kalogiannidis S, Melfou K (2020) Issues and opportunities for agriculture sector during global pandemic. Int J Econ Bus Manage Res 4(12):204–211

Khan I, Shah D, Shah SS (2021) COVID-19 pandemic and its positive impacts on environment: an updated review. Int J Environ Sci Technol 18(2):521–530

Laborde D, Martin W, Swinnen J, Vos R (2020) COVID-19 risks to global food security. Science 369(6503):500–502. https://doi.org/10.1126/science.abc4765

Lai CC, Shih TP, Ko WC et al (2020) Severe acute respiratory syndrome coronavirus 2 (SARS-CoV-2) and coronavirus disease-2019 (COVID-19): the epidemic and the challenges. Int J Antimicrob Agents 55:105924. https://doi.org/10.1016/j.ijantimicag.2020.105924

Mandal MAS, Kabir MJ, Kabir MS, Salam MA, Islam MA, Omar MI, Sarkar MAR, Rahman MC, Chawdhury A, Rahaman MS, Deb L, Aziz MA, Siddique MAB (2020) Harvesting of Boro paddy in Haor areas of Bangladesh: interplay of local and migrant labour, mechanized harvesters and COVID-19 vigilance in 2020. Bangladesh Rice Research Institute, Gazipur

Marlow S (2020) COVID-19: effects on the fertilizer industry. IHS Market 24(3):2–6

Morsy H, Salami A, Mukasa AN (2021) Opportunities amid COVID-19: advancing intra-African food integration. World Dev 139:105308

Muscogiuri G, Barrea L, Savastano S, Colao A (2020) Nutritional recommendations for CoVID-19 quarantine. Eur J Clin Nutr 74(6):850–851

Mustafa N (2020) Impact of the 2019–20 coronavirus pandemic on education. Int J Health Pref Res 4(1):25–30

Otmani A, Benchrif A, Tahri M, Bounakhla M, El Bouch M, Krombi MH (2020) Impact of Covid-19 lockdown on PM10, SO_2 and NO_2 concentrations in Salé City (Morocco). Sci Total Environ 735:139541

Otsuka K, Fan S (eds) (2021) Agricultural development: new perspectives in a changing world. International Food Policy Research Institute, Washington, DC. https://doi.org/10.2499/9780896293830

Owusu-Fordjour C, Koomson CK, Hanson D (2015) The impact of COVID-19 on learning: the perspective of the ghanaian student. Eur J Educ Stud 7:88–101

Peeri NC, Shrestha N, Rahman MS, Zaki R, Tan Z, Bibi S, Baghbanzadeh M, Aghamohammadi N, Zhang W, Haque U (2020) The SARS, MERS and novel coronavirus (COVID-19) epidemics, the newest and biggest global health threats: what lessons have we learned? Int J Epidemiol 49(3):717–726. https://doi.org/10.1093/ije/dyaa033

Phillipson J, Gorton M, Turner R, Shucksmith M, Aitken-McDermott K, Areal F (2020) The COVID-19 pandemic and its implications for rural economies. Sustainability 12:3975. https://doi.org/10.3390/su12103973

Prata JC, Silva ALP, Walker TR, Duarte AC, Rocha-Santos T (2020) COVID-19 pandemic repercussions on the use and management of plastics. Environ Sci Technol 54(13):7760–7765. https://doi.org/10.1021/acs.est.0c02178

Reardon T, Mishra A, Nuthalapati C, Bellemare M, Zilberman D (2020) Covid-19's disruption of India's transformed food supply chains. Econ Polit Wkly 55:18–22

Richards TJ, Rickard B (2020) COVID-19 impact on fruit and vegetable markets. Can J Agric Econ/revue Canadienne D'agroeconomie 68(2):189–194

Rizou M, Galanakis IM, Aldawoud TM, Galanakis CM (2020) Safety of foods, food supply chain and environment within the COVID-19 pandemic. Trends Food Sci Technol 102:293–299

Rockstrom J, Edenhofer O, Gaertner J, DeClerck F (2020) Planet-proofing the global food system. Nature Food 1:3–5. https://doi.org/10.1038/s43016-019-0010-4

Rolland RM, Parks SE, Hunt KE et al (2012) Evidence that ship noise increases stress in right whales. Proc R Soc B 279:2363–2368. https://doi.org/10.1098/rspb.2011.2429

Rowan NJ, Galanakis CM (2020) Unlocking challenges and opportunities presented by COVID-19 pandemic for cross-cutting disruption in agri-food and green deal innovations: Quo Vadis? Sci Total Environ 748:141362

Selvaranjan K, Navaratnam S, Rajeev P, Ravintherakumaran N (2021) Environmental challenges induced by extensive use of face masks during COVID-19: a review and potential solutions. Environ Challenges 3:100039. https://doi.org/10.1016/j.envc.2021.100039

Singhal T (2020) A review of coronavirus disease-2019 (COVID-19). Indian J Pediatr 87:281–286. https://doi.org/10.1007/s12098-020-03263-6

Stephens EC, Martin G, Van Wijk M, Timsina J, Snow V (2020a) Impacts of COVID-19 on agricultural and food systems worldwide and on progress to the sustainable development goals. Agric Syst 183:102873

Stephens EC, Martin G, van Wijk M, Timsina J, Snow V (2020b) Editorial: impacts of COVID-19 on agricultural and food systems worldwide and on progress to the sustainable development goals. Agric Syst 183:102873. https://doi.org/10.1016/j.agsy.2020.102873

Tiruneh D (2020) COVID-19 school closures may further widen the inequality gaps between the advantaged and the disadvantaged in Ethiopia. The Education and Development Form. https://www.ukfiet.org/2020/covid-19-school-closures-may-further-widen-the-inequality-gaps-between-the-advantaged-and-the-disadvantaged-in-ethiopia/

Torres FG, De-la-Torre GE (2021) Face mask waste generation and management during the COVID-19 pandemic: an overview and the Peruvian case. Sci Total Environ 786:147628. https://doi.org/10.1016/j.scitotenv.2021.147628

Tzifopoulos M (2020) In the shadow of coronavirus: distance education and digital literacy skills in Greece. Int J Social Sci Technol 5:1–14

UNESCAP (The United Nations Economic and Social Commission for Asia and the Pacific) (2020) COVID-19 and South Asia: national strategies and subregional cooperation for accelerating inclusive, sustainable and resilient recovery. Subregional Office for South and South-West Asia. https://www.unescap.org/sites/default/files/South%20Asia%20COVID-19%20Paper_5.pdf

UNESCO (2020b) UNESCO Rallies international organizations, civil society and private sector partners in a broad coalition to ensure #LearningNeverStops. UNESCO. https://en.unesco.org/news/unesco-rallies-international-organizations-civil-society-and-private-sector-partners-broad

Wang D, Hu B, Hu C et al (2020) Clinical characteristics of 138 hospitalized patients with 2019 novel coronavirus-infected pneumonia in Wuhan, China. JAMA J Am Med Assoc 323:1061–1069. https://doi.org/10.1001/jama.2020.1585

WHO—World Health Organization (2019) Air pollution. https://www9.who.int/airpollution/en/. Accessed 26 Dec 2022

Workie E, Mackolil J, Nyika J, Ramadas S (2020) Deciphering the impact of COVID-19 pandemic on food security, agriculture, and livelihoods: a review of the evidence from developing countries. Curr Res Environ Sustain 2:100014

World Bank (2020) Global economic prospects: a world bank group flagship report. https://www.worldbank.org/en/publication/global-economic-prospects

Xu Z, Ren W (2019) Application of a hybrid model based on echo state network and improved particle swarm optimization in PM2.5 concentration forecasting: a case study of Beijing, China. Sustainability. https://doi.org/10.3390/su11113096

Xu EG, Ren ZJ (2021) Preventing masks from becoming the next plastic problem. Front Environ Sci Eng 15(6):125. https://doi.org/10.1007/s11783-021-1413-7

Zahra FT, Saleem S, Imran M et al (2020) The SARS-CoV-2 pandemic and the role of honey and its products as an emerging therapeutic regime: a review. Biomedica 36:COVID 19-S2

Zowalaty MEE, Järhult JD (2020) From SARS to COVID-19: a previously unknown SARS-related coronavirus (SARS-CoV-2) of pandemic potential infecting humans-call for a one health approach. One Health. https://doi.org/10.1016/j.onehlt.2020.100124

Chapter 4
Interruptions in the Agricultural Production System and Food Value Chain in the Times Covid-19 Pandemic

Prisca Divra Johan, Osumanu Haruna Ahmed, Adiza Alhassan Musah, Nur Thaqifah Salihah Salleh, Mohd Hafiz Jamaludin, Rose Abdullah, Anisah Syakirah Anwari, and Zarina Zainuddin

Abstract The incidence and spread of the coronavirus pandemic (COVID-19) has created an imbalance in all sectors worldwide with a profound disruption on the global economy. Social distancing, quarantine regulations, and strict travel restrictions have led to a major reduction in the workforce and loss of jobs across all

---

P. D. Johan
Department of Crop Science, Faculty of Agricultural Science and Forestry, Universiti Putra Malaysia, Bintulu Sarawak Campus, 97008 Bintulu, Malaysia
e-mail: prisca.divra@gmail.com

O. H. Ahmed (✉) · N. T. S. Salleh · R. Abdullah · A. S. Anwari
Faculty of Agriculture, Universiti Islam Sultan Sharif Ali, KM 33, Jalan Tutong, Kampung Sinaut, Tutong TB1741, Brunei Darussalam
e-mail: ahmed.haruna@unissa.edu.bn

N. T. S. Salleh
e-mail: thaqifah.salleh@unissa.edu.bn

R. Abdullah
e-mail: rose.abdullah@unissa.edu.bn

A. S. Anwari
e-mail: anisah.anwari@unissa.edu.bn

A. A. Musah
Graduate School Management, Postgraduate Centre, Management and Science University, University Drive, Off Persiaran Olahraga Section 13, 40100 Shah Alam, Selangor, Malaysia
e-mail: adiza_alhassan@msu.edu.my

M. H. Jamaludin
Faculty of Agrobased Industry, Universiti Malaysia Kelantan, Jeli Campus, 17600 Jeli, Kelantan, Malaysia
e-mail: hafiz@umk.edu.my

Z. Zainuddin
Department of Plant Science, Kulliyyah of Science, International Islamic University Malaysia, 25200 Kuantan, Pahang Darul Makmur, Malaysia
e-mail: zzarina@iium.edu.my

D. Jones and A. J. Khan Chowdhury (eds.), *Environmental Resilience in the Pandemic Years 2020–2021*, Environmental Science and Engineering,
https://doi.org/10.1007/978-3-031-47757-7_4

industrial sectors. The agrarian sectors such as agriculture, fisheries, and livestock which are vital aspects of food systems have not been spared by COVID-19. The agricultural sector forms the backbone of the global economy as it provides livelihood to the developed and developing countries. Therefore, disruptions in food security and the agricultural sector is having and may continue to have far-reaching impacts on the agricultural ecosystem services. As a result, the pandemic has demonstrated the fragility of agriculture and the related food chain. Owing to the importance of the agrarian sectors, this review paper highlights the interruptions in the food and agriculture industry as a result of the on-going COVID-19 pandemic. The current state of the agriculture ecosystem in relation to the COVID-19 has indicated the importance of sustainable and resilient food system approaches to ensure future food security and food provision.

Keywords Coronavirus disease · Global pandemic · Lockdown · Agriculture sector · Food security · Sustainable development goals · Agricultural ecosystem · Supply chain

4.1 Introduction

The whole world is in a serious danger because of the novel coronavirus (COVID-19) pandemic and many countries worldwide have declared state of health emergency. On 11 March 2020, COVID-19 was declared as a global pandemic by the World Health Organization (WHO 2020a, b; Vasavada 2020). The first cases of the pandemic were identified in December 2019 when a pneumonia outbreak took place in the Wuhan City, Hubei Province, China (Abbott et al. 2020; Kumar et al. 2020). The emergence of COVID-19 rapidly spread across the world with the greatest mortality effects reported in the United States, Brazil, Mexico, India, the United Kingdom, Italy, and France (Worldometer 2021). The common signs and symptoms are dry cough, dyspnea, fever, tiredness, and lung trouble (Khan et al. 2021). Some patients, mostly the aged and with other chronic symptoms develop into pneumonia, with chest tightness, chest pain, and shortness of breath; however, some people may have no symptoms at all (Favalli et al. 2020). The COVID-19 spreads quickly in infestation and has a high mortality rate around the world (Khan et al. 2020). Quarantines and other restrictions have been implemented to combat pandemic and these measures are expected to remain in place for many weeks and months. Although vaccination may provide some mitigations many risks are still obvious and the impact on all sectors of economy is detrimental (Hörner et al. 2020; Patrinley et al. 2020).

Among many sectors, the food and the agriculture ecosystem was the worst hit (Hamid and Mir 2021). The COVID-19 poses numerous threats to the sustainability of the agricultural ecosystem because it is vulnerable due to food supply security requirements. The COVID-19 pandemic has also had many secondary damaging outcomes on the global sustainability of agricultural systems. The effects of the COVID-19 pandemic, for example, is glaring in the agricultural sector in terms of supply chain disruptions and agricultural production. When an infectious disease outbreak occurs, hunger and malnutrition increase (Burgui 2020; Sar et al. 2010).

As a result of the uncontrollable spread of COVID-19, countries imposed lockdowns, postponed or banned international travel, and limited exports and imports to control the transmission of COVID-19 (Sidor and Rzymski 2020; WHO 2020a, b). The governmental lockdown policy implementation led to labour scarcities during the harvest or obstacles for farmers bringing product to the market (Aday and Aday 2020; Gong et al. 2020). In addition, COVID-19 pandemic control measures delayed and interrupted transport and logistics services. The stagnation in the transportation system causing the products those produced on the farm are being wasted because it cannot be sold. The border closures and further procedures and investigations created congestion and delays, affecting the transport of perishable products (Pulubuhu et al. 2020). Furthermore, marginal farms suffered most or will continue to suffer because they are farmers are compelled to sell their products at nominal prices. As the price of food increases, on the other hand, farmers are struggling to recover the cost of their production.

Given the globalized nature of the COVID-19, there are concerns about the upcoming surge at all levels, thus it is obligatory to contemplate the significances on different scales in the form of dissimilar scenarios such that this serious crisis is sustainably managed and overcome. Agricultural sectors and food security may be limited, therefore rules and recommendations are needed to reduce the occurrence and contamination in agricultural societies and farms or farming systems. Keeping the threats of COVID-19 in human life, this paper reviews the virus disease's impacts on sustainability, with a focus on agriculture and food security in the developing world.

4.2 Food Security

Agriculture is one of the most important sectors in human development and it is critically tied to food security (Abdelhedi and Zouari 2020; Lopez-Ridaura et al. 2019). Food security refers to a situation where all individuals at all times have continuous physical and economic access to sufficient, safe, and nutritious food to fulfil their dietary needs and food choices for an active and healthy lifestyle (Elsahoryi et al. 2020). Food security has been jeopardized both directly and indirectly as a result of the virus's destabilization of food systems in addition to the effects of lockdowns on family revenue and physical access to food (Devereux et al. 2020). The incidence of COVID-19 has a negative impact on four dimensions of food security namely; availability of food, accessibility of food, utilization of food, and stability of food (Nechifor et al. 2021; Laborde et al. 2020). Some literature studied the impact of COVID-19 on all the food security dimensions (Guiné et al. 2021) and some studied only one or more (Jafri et al. 2021) dimensions. However, it is impossible to identify which food security dimensions is critically affected, unless the whole literature conducted on the impact of COVID-19 on food security is summarized.

According to Genkin and Mikheev (2020), the report by the Food and Agriculture Organization (FAO), World Trade Organization, and World Health Organization (WHO) highlights the threat of a food disaster triggered by the current COVID-19, with a risk of a global "food shortage" attributed to disruptions in the trade industry's supply chain. Prior to the emergence of COVID-19, approximately 820 million people were exposed to hunger daily, and more than two billion people faced micronutrient shortages each day and this exposes them to diseases and shortened life expectancy (Picchioni et al. 2021). Following the COVID-19 pandemic, food security is compromised because of restrictions in the production ecosystem, processing, and marketing of food. For example, fresh produce, particularly vegetables, fish, and meat, recorded high prices as a result of disruptions in distribution channels (Kansiime et al. 2021). Because of COVID-19, 3.3 billion global workforces are at risk of losing their livelihood (WHO 2020a, b). The food insecurity owing to COVID-19 affects the global society regardless of their economic status (Niles et al. 2021; Kent et al. 2022), but severity of the food insecurity is substantial in lower income countries than in the developed world (Gebeyehu et al. 2022).

4.3 Impacts of COVID-19 on Agricultural Systems

Ancog and Gregorio (2020) highlighted that the COVID-19 pandemic creates supply and demand shocks, affecting all economic sectors, including agriculture. These two aspects are directly related to food security, thus they might lead to food security at risk. A study by Tripathi et al. (2021) revealed that COVID-19 has caused the following challenges: (i) income losses and falling demands, (ii) supply chain breakdown, (iii) changed consumer reactions and dietary changes, and (iv) international and domestic trade disruptions. Considering the food supply chain, it is obvious that COVID-19 has an impact on the whole process which disconnects farm production to the consumer. Work from home had been recommended for various sectors but this recommendation cannot be applied to the agricultural and food sectors where daily routine activities need to continue unabated. Road closures and transport restrictions retarded or impeded agricultural services, and this has led to loss of produce and income. In addition, the agricultural sector which depends on foreign workers is facing labour acute shortage. This affects crops, livestock, and fishery sectors such that the impact is felt along the whole agriculture ecosystem and its value chain, hence affecting food availability and market prices for that matter.

4.3.1 Crops

The scale of the incidence COVID-19 compelled many countries to impose lockdown several times and the primary motive was control or contain the rapid spread of this disease. The lockdown restricts movement of farming goods and communities, especially labourers or farm workers who must generally move from one place to another to work and earn income. A recent report estimated that there has been an overall decrease in agriculture production in Southeast Asian countries by 3.11% (17.03 million tons) in the first quarter of 2020 because of the absence of labourers or farm workers (Ancog and Gregorio 2020). Moreover, restrictions on movement hampered the smooth flow of agricultural inputs such as fertilizers, pesticides, seeds, machinery spare parts, and farm equipment. This disrupted the agriculture production system and the associated food supply chain disruption (Barrett 2020; Aromolaran and Muyanga 2020). In addition, farmers had difficulties in marketing their products. Perishable goods such as fresh fruits and vegetables are the most affected areas of the agriculture production system. This resulted in increased levels of post-harvest lost (PHL) and food wastes production at an alarming rate. Agricultural food losses are primarily associated with the spoilage of a portion of the avoidable environmental effects of food supply chains, and are caused by the perishability of agricultural produce, transportation issues, and the inherent difficulty in achieving demand and supply equilibrium (Adelodun and Choi 2020). Furthermore, during the COVID-19 pandemic, consumers' food preferences shifted, increasing demand for specific type of farm produce. For example, the consumer paid special attention to vitamin C rich fruits because it can boost up the immunity system.

4.3.2 Livestock

The impact of COVID-19 on the livestock industry is mostly unknown and undefined. Although formal assessments are not yet attainable, recent findings suggest that the livestock supply chain has been disrupted (Haque et al. 2022). During the first phase of the Movement Control Order (MCO), some small-scale farmers and distributors had difficulty selling their products because of bad market access and logistical constraints. A similar scenario occurred in the dairy and poultry industry (Vaghefi 2020). Because the maximum amounts of raw materials for feed are imported, an interruption in the import–export business created a crisis of poultry feed. Suppliers reported that disruption in the global trade meant they could not import component raw materials to produce feed and medicine (Sattar et al. 2021). This is supported by reports of prolonged border closures preventing the import of essential feed items such as maize. Also, production in the pharmaceutical industries was severely disrupted due to blockade in the importation of Active Pharmaceutical Ingredients (API) (IDLC 2020). This resulted in inadequate veterinary services. Furthermore, the poultry sector was affected by the spread of misinformation that COVID-19 can be contracted through eggs and chickens. Hence, the decrease in demand for eggs and chickens. Because of lack of storage facilities, thousands of gallons of milk, eggs, vegetables, and fruit were wasted. Different super shops, dairy food outlets, and restaurants remain closed during the lockdown period. As a result, the daily consumption rate of milk, eggs, and meat decreased.

4.3.3 *Fishery*

In many cases within different developing countries, COVID-19 has severely affected the local and commercial aquatic food production systems, both directly and indirectly. Because of the significant decrease in the market's demand for fish and the limited transportation options that were available during the lockdown, fish farms had difficulty in collecting and selling their goods (FAO 2021; Sunny et al. 2021). Many wholesalers also stopped buying fish and aquatic products because restaurants were ordered to remain shut. In areas of high fish production, unsold fish was consumed in order to support food security. Farmers' inability to sell their products has resulted in an increase in live fish stock levels and a lengthening of the fish culture period, both of which have harmed feed conversion ratios, the ability to restock, and, ultimately, farm profitability (Alam et al. 2022). Accordingly, feeding expenditures and the risk of fish mortality have increased. There were no ways or channels to start a new farming cycle because the farmers were unable to harvest their current produce. As a result, they must attempt to sell their produce at a lower price, which results in negative economic consequences for the farmers' livelihoods (FAO 2020; Sunny et al. 2021). Additionally, many employees who were working in the processing, harvesting, and marketing of aquatic food have lost their jobs (Bennett et al. 2020). This trend is therefore expected to have a significant economic effect on the fishing industry and fishing communities.

4.4 SWOT Analysis of COVID-19 Prevention and Strategies

SWOT analysis refers to the assessment of innumerable strengths (S), weaknesses (W), opportunities (O), threats (T), and other factors concerning the topic of interest. In this context, the SWOT analysis covers the impact of COVID-19 on agricultural food supply, demand, and insecurity.

4.4.1 *Strengths*

The lockdowns across countries compelled the use of information systems from 40 to 100% (Pandey and Pal 2020). In many countries, smartphone applications are available and can be strengthened to connect producers and buyers (Amjath-Babu et al. 2020). Many applications enable farmers to sell their products with one click. The use of mobile wallet also saw an increase during the pandemic in many developing countries (FAO 2021a, b; Mehrolia et al. 2020). One of the most important aspects a mobile application offers is the creation of target market and the potential for fostering trust between producers and consumers. Moreover, the virtual interface has promoted diverse and vibrant farmers and consumers. For example, fresh fruits and vegetables can be sold from the farm and directly shipped to end customers. Farmers can utilize social media accounts to post videos of fresh produce and advertise their products to groups of potential customers. Thus, through social media platforms, farmers connect with consumers individually and build trust by switching from traditional shops to online buying.

4.4.2 Weaknesses

During the pandemic, there was a lack of transport for the mobility of products between provinces, states, regions, or districts and even countries and continents. In this regard, e-commerce platforms could facilitate cooperation between farmers and local or international consumers to boost online sales. The development of e-commerce systems in developing countries has the potential to create an end-to-end network connecting rural supply and urban demand with a possibility of even going global. This would address the current issue of agribusiness stakeholders being excluded from decision-making or having no authority to influence government decisions. This challenge may be partially attributed to the steep learning curve required to learn and use technologies such as video and audio-conferencing platforms to arrange remote meetings between policymakers and agribusiness stakeholders. Therefore, it would be beneficial if not essential to invest time in providing tutorials and technological assistance for farmers such that they can utilize e-commerce platforms to their full potential. Moreover, inadequate or lack of a marketing, communication strategy, and funding for farmers to grow crops are major reasons for the disruption of food supply around the globe. Hence, farmers should be given targeted subsidies or interest-free loans to ensure long-term production.

4.4.3 Opportunities

The implementation of control measures across countries has revealed issues with product mobility. With precautionary measures, the governments are easing COVID-19 restrictions to open small markets and allied industries across the countries in addition to incorporating digital technologies to maintain daily life and economic activities. Countries can provide farmers with counselling and agricultural extension services (agriculture, socio-economic, and environmental efficiency) through digital platforms. In this manner, digital transformation can lower transaction costs, improve supply chain coordination, and benefit many farmers and consumers (Abid et al. 2020). Transformation of the traditional supply chain into digital supply networks would enable dealing with market uncertainty and supply chain problems in future (Zhu et al. 2020). Digitizing the relationship between the buyer and the producer is also a core aspect for developing robust supply chains. By managing all aspects of the supply chain digitally with transparency of information sharing can effectively allow agribusinesses to handle bullwhip effect in future. Besides, digital platforms can ensure growing public interest in transparent pricing and encourage virtual trade instead of currency usage.

4.4.4 Threats

During the COVID-19 outbreak, the emergence of new regulations increased trade barriers and disrupted the agricultural production system and the accompanied food supply chain. Some agribusinesses are shutting down because of bankruptcy and the loss of perishable food products. The sudden lockdown caused massive disruptions in the agricultural production system and food supply chains because of the significant decrease demand. As the pandemic spreads, this decrease in demand resulted in significant amounts of food waste and this compelled growers of perishable products to dumped crops into the ground because of labour shortages. As a result of the COVID-19 outbreak, the agricultural economic activities contracted globally. A sharp decrease in buying power and a high rate of unmet financial need resulted in suspension of food delivery. In these times of crises, food banks, supported by private charitable organizations, should be mobilized to distribute food, to families in need.

4.5 Recommendations to Minimize the Effects of COVID-19 on Agricultural Production Systems

The COVID-19 outbreak seriously threatens food safety, security, and nutrition. Disrupted marketing, logistics and commercial networks, and potential problems could limit access to food in some parts of the world (FAO 2020). Report from World Food Programme demonstrated that the number of people facing extreme hunger can increase to 265 million in 2020 as a result of COVID-19 (WFP 2020). Research undertaken by Headey et al. (2020) found that COVID-19 contributes to increase of 14.3% in the incidence of lack of health and social security for low or middle-income children under five years of age.

4.5.1 Suggestions for Small Farmers

Countries should take measures to ensure the safety of agricultural workers. Employees' illness status should be tracked by on-site healthcare professionals. Countries should construct agricultural production collection centres in easily accessible locations for farmers in order to reduce mobility. Furthermore, agricultural production collection centres should be designed to provide high-capacity storage. Improved and advanced storage structures can also be used to minimize the loss of foods throughout the food value chain. Food banks can play an important role in the horizontal and vertical coordination mechanisms with farmer associations that make contractual agriculture arrangements. Primarily, it would be a way to assist farmers in creating new markets by selling unsold produce to food banks and connecting farmers with vulnerable people during the COVID-19 outbreak (Jackson and Yurkevich 2020). Besides, countries should participate in growth and rapid development of e-commerce for small shareholders. Communication through the internet ensures that commercializing produces to wider range of consumers and enables farmers to find cheaper inputs (Khanal and Mishra 2016). In addition, small-scale producers should have easy and unhindered access to credit for dealing with financial problems to continue production. Some countries offer incentive packages for small-scale farmers. Trade restrictions and bureaucratic barriers should be removed to ensure that small-scale farmers and producers have access to markets. Governmental agencies must meet the energy requirements of small-scale producers in rural areas. Crop diversification strategy is also important to increase the diversity of food by crop rotation or intercropping for creating new marketing channels and enabling the harvesting throughout the year can be used to adapt the challenges easily (Hufnagel et al. 2020). Buyer, investors, and bankers should pay the small farmers upfront for their produces to ensure liquidity for next season. Growing organic foods can also scale up the sales due to consumer interest and confidence. Additionally, partnerships between small farmers and companies or government is another way to help the small farmers to increase productivity and incomes by adaptive technology investment (Winter and Davis 2006).

4.5.2 Suggestions for Government and Business

A crisis committee should be established to focus on the impact of COVID-19 on the food value chain without delaying the implementation of specific strategies and interventions for too long. This committee should play an important role in monitoring progress and recommending actions to mitigate the effects of COVID-19 on agricultural production and food supply cuts. Furthermore, governments should develop and implement emergency provisioning strategies to support production. Temporary input subsidy programs should be implemented in the areas most affected by the outbreak. Timely assistance is critical for the upcoming planting season. Apart from this, migrant data collection and assessment programmes should be used to determine when and where migrants are needed. Facilitating migrant workers' cross-border movement is important because movement restrictions and border closures have a significant negative impact on agricultural labour supply. Online platforms should be used to facilitate connections between local residents and agriculture sector. Unemployed people or local workers should be encouraged to work in agriculture by adding premiums to wages. This is because public and private standards are used to define the minimum requirements in food safety and quality, employment contracts made between actors in the food value chain should be fair to all parties and clear about the rights and responsibilities of the parties. Furthermore, logistic operations are critical to maintaining the delivery of the foods, and, therefore, some efforts are needed to maintain operations. More supervision services, upgraded sanitation systems, increased use of digital documentation, and operations will necessitate infrastructure investment. Rapid yield prediction and determination of national food stocks are required to define any shortages or surpluses that may occur, particularly because of import prohibition or export restrictions. Crop yield information models can be used to assist governments in making decisions about food security or grain marketing. To understand the consequences of policy decisions, a suitable modelling method should be chosen.

4.6 Conclusion

The agricultural sector is one of the most directly affected by the movement's control order because of COVID-19 pandemic outbreaks. Farmers have experienced several obstacles owing to COVID-19 such as input supply limitation, lack of technical support, inability to market their products, lack of transportation to the market, export restrictions on agriculture products, and low prices. These difficulties have resulted in inadequate output, unanticipated stock retention, loss of returns, and food insecurity. Many COVID-19-related restrictions have contributed to a significant gap between demand and supply in the food chain, resulting in a decline of the agricultural industry. Food supply chain disruptions highlight the importance of risk management. The agriculture sector needs to be more resilient to production and market risks. On-farm and market-based risk management strategies and tools can help farmers cope with uncertainties. In the long term, countries needs to invest in disaster preparedness, especially where the food supply chain is concerned. It is important to understand that certain precautionary and control efforts compromise agricultural production.

Institutional Review Board Statement Not applicable.

Informed Consent Statement Not applicable.

Data Availability Statement The study did not report any data.

Acknowledgements The authors would like to acknowledge Universiti Putra Malaysia, Universiti Islam Sultan Sharif Ali, Management and Science University, Universiti Malaysia Kelantan, and International Islamic University Malaysia for this review paper.

Conflict of Interest The authors confirm that they have no conflict of interest to declare for this publication.

References

Abbott S, Hellewell J, Munday J, Funk S, CMMID nCoV Working Group (2020) The transmissibility of novel Coronavirus in the early stages of the 2019–20 outbreak in Wuhan: exploring initial point-source exposure sizes and durations using scenario analysis. Wellcome Open Res 5:17

Abdelhedi IT, Zouari SZ (2020) Agriculture and food security in North Africa: a theoretical and empirical approach. J Knowl Econ 11(1):193–210

Abid A, Jie S, Aslam W, Batool S, Lili Y (2020) Application of struc-tural equation modelling to develop a conceptual model for smallholder's credit access: the mediation of agility and innovativeness in organic food value chain finance. PLoS ONE 15(8):e0235921

Aday S, Aday MS (2020) Impact of COVID-19 on the food supply chain. Food Quality Safety 4(4):167–180

Adelodun B, Choi KS (2020) Impact of food wastage on water resources and GHG emissions in Korea: a trend-based prediction modeling study. J Clean Prod 271:122562

Alam GMM, Sarker MNI, Gatto M, Bhandari H, Naziri D (2022) Impacts of COVID-19 on the fisheries and aquaculture sector in developing countries and ways forward. Sustainability 14(3):1071

Amjath-Babu TS, Krupnik TJ, Thilsted SH, McDonald AJ (2020) Key indicators for monitoring food system disruptions caused by the COVID-19 pandemic: insights from Bangladesh towards effective response. Food Security 12(4):761–768

Ancog RC, Gregorio GB (2020) Impact of COVID-19 pandemic on agriculture production in Southeast Asia: reinforcing transformative change in agricultural food systems (no. 2020–1). Southeast Asian Regional Center for Graduate Study and Research in Agriculture (SEARCA)

Aromolaran A, Muyanga M (2020) Impact of COVID-19 on food systems and rural livelihoods in Nigeria–round 1 report

Barrett D (2020) Dreams about COVID-19 versus normative dreams: trends by gender. Dreaming 30(3):216

Bennett NJ, Finkbeiner EM, Ban NC, Belhabib D, Jupiter SD, Kittinger JN, Mangubhai S, Scholtens J, Gill D, Christie P (2020) The COVID-19 pandemic, small-scale fisheries and coastal fishing communities. Coast Manag 48(4):336–347

Burgui D (2020) Coronavirus: how action against hunger is responding to the pandemic. https://www.actionagainsthunger.org/story/coronavirushow-action-against-hunger-respondingpandemic

Devereux S, Béné C, Hoddinott J (2020) Conceptualising COVID-19's impacts on household food security. Food Security 12(4):769–772

Elsahoryi N, Al-Sayyed H, Odeh M, McGrattan A, Hammad F (2020) Effect of Covid-19 on food security: a cross-sectional survey. Clin Nutrit ESPEN 40:171–178

FAO (2020) Summary of the Impacts of the COVID-19 pandemic on the fisheries and acquaculture sector. FAO, Rome. ISBN 978-92-5-132789-0

FAO (2021a) Digital finance and inclusion in the time of COVID-19. Lessons, experiences, and proposals. Food and Agriculture Organization. http://www.fao.org/3/cb2109en/CB2109EN.pdf

FAO (2021b) The impact of COVID-19 on fisheries and aquaculture food systems, possible responses. FAO: Food and Agriculture Organization of the United Nations, Rome. ISBN 978-92-5-133768-4

Favalli EG, Ingegnoli F, De Lucia O, Cincinelli G, Cimaz R, Caporali R (2020) COVID-19 infection and rheumatoid arthritis: Faraway, so close! Autoimmun Rev 19(5):102523

Gebeyehu DT, East L, Wark S, Islam MS (2022) Impact of COVID-19 on the food security and identifying the compromised food security dimension: a systematic review protocol. PLoS ONE 17(8):e0272859

Genkin A, Mikheev A (2020) Influence of coronavirus crisis on food industry economy. Foods Raw Mater 8(2):204–215

Gong H, Hassink R, Tan J, Huang D (2020) Regional resilience in times of a pandemic crisis: the case of COVID-19 in China. Tijdschr Econ Soc Geogr 111(3):497–512

Guiné RDPF, Pato MLDJ, Costa CAD, Costa DDVTAD, Silva PBCD, Martinho VJPD (2021) Food security and sustainability: discussing the four pillars to encompass other dimensions. Foods 10(11):2732

Hamid S, Mir MY (2021) Global agri-food sector: challenges and opportunities in covid-19 pandemic. Front Sociol 142

Haque M, Zaman M, Uz R, Rahman M, Hossain M, Shurid TI, Rimi TA, Arby H, Rabbany M (2022) A review on impacts of COVID-19 on global agricultural system and Scope for Bangladesh after pandemic. Environ Sci Pollut Res 29:1–12

Headey D, Heidkamp R, Osendarp S, Ruel M, Scott N, Black R et al (2020) Impacts of COVID-19 on childhood malnutrition and nutrition-related mortality. Lancet 396:519–521

Hörner C, Schürmann C, Auste A, Ebenig A, Muraleedharan S, Dinnon KH III, Scholz T, Herrmann M, Schnierle BS, Baric RS, Mühlebach MD (2020) A highly immunogenic and effective measles virus-based Th1-biased COVID-19 vaccine. Proc Natl Acad Sci 117(51):32657–32666

Hufnagel J, Reckling M, Ewert F (2020) Diverse approaches to crop diversification in agricultural research. a review. Agron Sustain Dev 40(2):1–17

IDLC (2020). Feed industry of Bangladesh: sustaining COVID-19 and potentials in upcoming days. In: IDLC monthly business review. https://idlc.com/mbr/images/public/xWV4Ylp7Dg1TyTKLENXGmW.pdf

Jackson A, Yurkevich V (2020) Farmers are throwing out food that could go to food banks. American Farm Bureau and Feeding America want to change that, CNN

Jafri A, Mathe N, Aglago EK, Konyole SO, Ouedraogo M, Audain K, Zongo U, Laar AK, Johnson J, Sanou D (2021) Food availability, accessibility and dietary practices during the COVID-19 pandemic: a multi-country survey. Public Health Nutr 24(7):1798–1805

Kansiime MK, Tambo JA, Mugambi I, Bundi M, Kara A, Owuor C (2021) COVID-19 implications on household income and food security in Kenya and Uganda: findings from a rapid assessment. World Dev 137:105199

Kent K, Murray S, Penrose B, Auckland S, Horton E, Lester E, Visentin D (2022) The new normal for food insecurity? A repeated cross-sectional survey over 1 year during the COVID-19 pandemic in Australia. Int J Behav Nutr Phys Act 19(1):1–13

Khan KS, Mamun MA, Griffiths MD, Ullah I (2020) The mental health impact of the COVID-19 pandemic across different cohorts. Int J Mental Health Addict [Epub ahead of print]

Khan MU, Ali SZ, Habib K, Khan H, Tariq F, Bibi S (2021) A novel intelligent model for COVID-19 detection using cough auscultations and Hjorth descriptors. In: ICAME21, international conference on advances in mechanical engineering, Pakistan, pp 1–10

Khanal AR, Mishra AK (2016) Financial performance of small farm business households: the role of internet. China Agric Econ Rev 8(4):553–571

Kumar D, Malviya R, Sharma PK (2020) Corona virus: a review of COVID-19. EJMO 4(1):8–25

Laborde D, Martin W, Swinnen J, Vos R (2020) COVID-19 risks to global food security. Science 369(6503):500–502

Lopez-Ridaura S, Barba-Escoto L, Reyna C, Hellin J, Gerard B, van Wijk M (2019) Food security and agriculture in the Western Highlands of Guatemala. Food Security 11(4):817–833

Mehrolia S, Alagarsamy S, Solaikutty VM (2020) Customers response to online food delivery services during COVID-19 outbreak using binary logistic regression. Int J Consum Stud 45(3):1–13

Nechifor V, Ramos MP, Ferrari E, Laichena J, Kihiu E, Omanyo D, Musamali R, Kiriga B (2021) Food security and welfare changes under COVID-19 in Sub-Saharan Africa: impacts and responses in Kenya. Glob Food Sec 28:100514

Niles MT, Wirkkala KB, Belarmino EH, Bertmann F (2021) Home food procurement impacts food security and diet quality during COVID-19. BMC Public Health 21(1):1–15

Pandey N, Pal A (2020) Impact of digital surge during Covid-19 pandemic: a viewpoint on research and practice. Int J Inf Manage 55:102171

Patrinley JR, Berkowitz ST, Zakria D, Totten DJ, Kurtulus M, Drolet BC (2020) Lessons from operations management to combat the COVID-19 pandemic. J Med Syst 44(7):129

Picchioni F, Goulao LF, Roberfroid D (2021) The impact of COVID-19 on diet quality, food security and nutrition in low- and middle-income countries: a systematic review of the evidence. Clin Nutrit

Pulubuhu DAT, Unde AA, Sumartias S, Sudarmo S, Seniwati S (2020) The economic impact of COVID-19 outbreak on the agriculture sector. Int J Agric Syst 8(1):57–63

Sar TT, Aernan PT, Houmsou RS (2010) H1N1 influenza epidemic: public health implications for Nigeria. Int J Virol 6:1–6

Sattar AA, Mahmud R, Mohsin MAS, Chisty NN, Uddin MH, Irin N, Barnett T, Fournie G, Houghton E, Hoque MA (2021) COVID-19 impact on poultry production and distribution networks in Bangladesh. Front Sust Food Syst 5:714649

Sidor A, Rzymski P (2020) Dietary choices and habits during COVID-19 lockdown: experience from Poland. Nutrients 12(6):1657

Sunny AR, Sazzad SA, Prodhan SH, Ashrafuzzaman M, Datta GC, Sarker AK, Rahman M, Mithun MH (2021) Assessing impacts of COVID-19 on aquatic food system and small-scale fisheries in Bangladesh. Mar Policy 126:104422

Tripathi HG, Smith HE, Sait SM, Sallu SM, Whitfield S, Jankielsohn A, Kunin WE, Mazibuko N, Nyhodo B (2021) Impacts of COVID-19 on diverse farm systems in Tanzania and South Africa. Sustainability 13(17):9863

Vaghefi N (2020) The heavy impact of Covid-19 on the agriculture sector and the food supply chain. Covid-19 Crisis Assess 19

Vasavada P (2020) COVID-19 and the food industry: what we know. Food Quality and Safety. https://www.foodqualityandsafety.com/article/covid-19-and-the-food-industry-what-we-know/

WFP (World Food Programme) (2020) COVID-19 will double number of people facing food crises unless swift action is taken. https://www.wfp.org/news/covid-19-will-double-number-people-facing-foodcrises-unless-swift-action-taken

Winter CK, Davis SF (2006) Organic foods. J Food Sci 71(9):R117–R124

World Health Organization (2020a) Coronavirus disease (COVID-19)

World Health Organization (2020b) Impact of COVID-19 on people's livelihoods, their health and our food systems. https://www.who.int/news/item/13-10-2020-impact-of-covid-19-on-people's-livelihoods-their-health-and-our-food-systems

Worldometer (2021) COVID-19: coronavirus pandemic. https://www.worldometers.info/coronavirus/. Accessed 17 April 2021

Zhu G, Chou MC, Tsai CW (2020) Lessons learned from the COVID-19 pandemic exposing the shortcomings of current supply chain operations: along-term prescriptive offering. Sustainability 12(14):5858

Chapter 5
The Effect of Environmental Factors on Covid-19: A Systematic Review

Nazri Che Dom, Norezuan Abdul Gani, Mohd Shukri Mohd Aris, Samsuri Abdullah, Hasber Salim, Sharifah Norkhadijah Syed Ismail, and Nopadol Precha

Abstract As a result of its global spread, COVID-19 is currently regarded as a pandemic. Whether or not climatic factors have an effect on COVID-19 transmission and diffusion is a crucial aspect. On the basis of the significance of relationships between environmental factors (humidity, temperature, rainfall, population density, and speed of wind) and COVID-19, this research offers a thorough evaluation and

---

Authors' Contributions This collaboration work was carried out among all the authors. Muhammad Afiq Brohan, Nazri Che Dom, Nopadol Precha, Hasber Salim and Sharifah Norkhadijah Syed Ismail designed the study; Muhammad Afiq Brohan and Nazri Che Dom collected the data and performed the analysis. Mohd Shukri Mohd Aris and Samsuri Abdullah interpreted the data and wrote the manuscript. All authors read and approved the final manuscript.

N. C. Dom · N. A. Gani · M. S. M. Aris (✉)
Centre of Environmental Health and Safety Studies, Faculty of Health Sciences, Universiti Teknologi MARA (UiTM), UITM Cawangan Selangor, 42300 Puncak Alam, Selangor, Malaysia
e-mail: myshukri@uitm.edu.my

N. C. Dom
e-mail: nazricd@uitm.edu.my

N. A. Gani
e-mail: 2021427382@student.uitm.edu.my

N. C. Dom
Integrated Mosquito Research Group (I-MeRGe), Universiti Teknologi MARA (UiTM), UITM Cawangan Selangor, 42300 Puncak Alam, Selangor, Malaysia

Institute for Biodiversity and Sustainable Development (IBSD), Universiti Teknologi MARA, 40450 Shah Alam, Selangor, Malaysia

M. S. M. Aris
Occupational Health and Safety Risk Management (OHSeRM) Research Initiative Group and Centre for Environmental Health and Safety, Universiti Teknologi MARA (UiTM), UITM Cawangan Selangor, 42300 Puncak Alam, Selangor, Malaysia

S. Abdullah
Faculty of Ocean Engineering Technology and Informatics, Universiti Malaysia Terengganu, 21030 Kuala Nerus, Terengganu, Malaysia
e-mail: samsuri@umt.edu.my

D. Jones and A. J. Khan Chowdhury (eds.), *Environmental Resilience in the Pandemic Years 2020–2021*, Environmental Science and Engineering,
https://doi.org/10.1007/978-3-031-47757-7_5

meta-analysis. Review Manager (RevMan) version 5.4 was used to conduct the meta-analysis. The study's 16 articles have conducted in-depth analysis on over 110 datasets that are country-specific. We noted significant correlations between temperature (0.31 [0.26, 0.39] for fixed effect incidence), humidity (0.64 [0.55, 0.75] for fixed effect incidence), wind speed (0.50 [0.32, 0.77] for fixed effect incidence), and population density (1.53 [1.34, 1.75] for fixed effect incidence) and the incidence of COVID-19 ($p < 0.01$). As a result, we can show that environmental aspects may be important elements of COVID-19.

Keywords Climate · COVID-19 · Environmental factor · Humidity · Temperature

5.1 Introduction

Over a million people have died as a result of COVID-19 as of 1 October 2020. However, many SARS-CoV-2 infected individuals still have the virus in their upper respiratory tracts. In spite of a viral RNA PCR positive status, individuals do not show any clear symptoms (Huang et al. 2020; Zhu et al. 2020). Besides direct, indirect, or aerosol contact, environmental aspects have had a significant impact on coronavirus transmission and survival (Liu 2020). Environment-related illness progression has not garnered much attention so far.

The transmission of infectious diseases is influenced by climatic factors (temperature and humidity), density of populace, and the standard of medical care (Wang et al. 2021). Tosepu et al. (2020) found a positive relationship between COVID-19 and temperature. In other words, the lowest average temperature of 26.1 °C and the highest average temperature of 28.6 °C are strongly correlated. There is a correlation between long-term temperature data and the prevalence of infections in low-altitude Indian regions that experience dry and hot weather. These locations are more likely to have COVID-19 transmission (Gupta et al. 2020). In 30 Chinese provinces, the COVID-19 infection rate was influenced by the weather. Hubei province recorded the highest rate of COVID-19 infections, with cases decreasing by 11–22% for every

H. Salim
School of Biological Sciences, Universiti Sains Malaysia, 11800 Penang, Malaysia
e-mail: hasbersalim@usm.my

S. N. S. Ismail
Department of Occupational Health and Safety, Faculty of Medicine and Health Sciences, Universiti Putra Malaysia, 43400 Serdang, Selangor, Malaysia
e-mail: norkhadijah@upm.edu.my

N. Precha
Department of Environmental Health and Technology, School of Public Health, Walailak University, Nakhon Si Thammarat, Thailand
e-mail: nopadol.pr@wu.ac.th

1 °C increase in relative humidity and by 36–57% for every 1 °C increase in temperature. In other provinces, these aspects were not constant (Qi et al. 2020). Additionally, in Turkey (Şahin 2020), Brazil (Prata et al. 2020), Mexico (Mendez-Ariaga), and the United States (Sehra et al. 2020), COVID-19 transmission was less frequent in hotter climates. In Australia, a similar connection with humidity was noted; however, viral transmission was unaffected by temperature (Ward et al. 2020).

Over time, there is a weakening correlation between the COVID-19 epidemic and relative humidity. A substantial link between coronavirus infections and a number of metrological markers has been discovered in earlier studies (Tan et al. 2005; Rabenau et al. 2005; Sajadi et al. 2020). The majority of studies show a positive association between humidity and COVID-19 occurrence (Haque and Rahman 2020; Rashed et al. 2020; Sharma et al. 2020; Sharif and Dey 2021; Pani et al. 2020). Coronavirus spread is influenced by the environment (humidity, temperature, and wind speed), density of populace, and accessibility to medical services. There is a need to comprehend the weather (humidity, temperature, and wind speed) (Dalziel et al. 2018). The dearth of population immunity, according to climate-reliant epidemic modelling, is a considerably more important component when it comes to virus transmission. Furthermore, in spite of warmer temperatures, the COVID-19 dissemination is not considerably slowed down by the summertime climate (Baker et al. 2020). Although there is a wealth of material available, our primary objective was to determine how environmental aspects impact COVID-19 cases and other human coronaviruses. We selected Asian nations since they are home to half of the world's populace.

5.2 Methods

The impact of weather on COVID-19 mortality and incidence rates is being investigated for the first time in this study. We began by looking through digital archives for informative articles and worthwhile data. We carried out a thorough systematic investigation using appropriate inclusion and exclusion norms. We evaluated the danger levels associated with each article individually as part of this study, verified the veracity of the evidence to see whether the articles were appropriate for a thorough review of the weather and COVID-19, and gathered useful information. Five weather factors—humidity, temperature, UV light, rainfall, and wind speed—were examined for associations with the COVID-19 fatality rate. Then, to evaluate publication bias and heterogeneity, which were quite challenging, we utilised forest and funnel plots.

This segment goes over how to find publications about COVID-19 instances and environmental influences. Three scientific databases, Scopus, Web of Science (WoS), and PubMed, were used to find the articles. The study's resources, inclusion and exclusion requirements, review methods (identification, screening, eligibility), and data abstraction and analysis were all modified in accordance with the PRISMA standards. Environmental aspects, climate change, and COVID-19 were the three main search terms used. People with COVID-19 diagnoses made up the population. Comparisons were made between nations, study parameters, climate variables, COVID-19 results, and qualitative research technique.

Scopus, Web of Science (WoS), and PubMed are the three main journal databases upon which the review and analysis procedures are based. WoS is a sizable database that includes more than 33,000 publications from over 256 different fields, such as multidisciplinary social sciences, environmental studies, social issues, and development and planning. It includes nearly a century's worth of comprehensive backfile and citation data gathered by Clarivate Analytics and grades them by utilising three metrics—papers, citations, and citations per paper. The second database considered for this review is Scopus. With 22,800 papers from 5000 publishers worldwide, it is one of the largest collections of abstracts and citations for peer-reviewed works. Scopus includes social, environmental, agricultural, and biological sciences among its many fields of study. Over 33 million citations and biomedical literature abstracts can be found in the PubMed database. Although links to content from other websites, such as publisher websites or PubMed Central (PMC), are frequently given, full-text journal articles are not included.

There were many set eligibility and exclusion norms. The search efforts did not include non-English sources and focused solely on English-language content to minimise ambiguity or difficulty in translation. Second, only publications from peer-reviewed journals were used to collect empirical data; book series, individual books, chapters, review articles, and conference proceedings were excluded. Third, since the majority of the pieces were released in 2019, there were no time restrictions regarding chronology. The analysis highlighted the relationship between COVID-19 instances and environmental factors (humidity, temperature, rainfall, population density, and vegetation area). Studies that looked at the relationship between environmental aspects and COVID-19 recovery and relapse rates were not included. Lastly, to further the goal of the research, we chose publications about Asian regions (Table 5.1).

The systematic review process was divided into these four sections: search terms were identified in the first round. To find words that were comparable to and connected to climate change, environmental aspects, and COVID-19, earlier studies and thesaurus were examined. At this point, careful screening assisted in removing

Table 5.1 Inclusion and exclusion criteria for the systematic reviews

Criteria	Eligibility	Exclusion
Language	English	Non-english
Type of literature	Journal (research articles)	Book, book series, chapter in book, conference proceeding
Correlation between environmental factors	Temperature, humidity, population density, rainfall and vegetation area	Recovery rates and reproduction rates
Countries	Asian countries	Non-Asian countries

685 duplicate items. The second phase, screening, involved removing 1044 items from the 1096 review-eligible papers. The papers were extensively evaluated during the third step to ascertain eligibility. A total of 52 papers were scrutinised carefully, but they were left out because they did not emphasise environmental aspects, were not empirical, or did not concentrate on Asian nations and regions. The last assessment step resulted in 16 articles for qualitative analysis (Fig. 5.1).

We used the Joanna Briggs Institute (JBI) tool29 to evaluate the papers we chose for this investigation. The checklist included eight questions, comprising (a) were the criteria for inclusion in the sample clearly outlined; (b) were the study participants and the setting outlined in detail; (c) was the exposure gauged in a valid and trustworthy manner; (d) were objective, standard norms utilised for gauging the condition; (e) were confounding aspects identified; (f) were strategies to tackle confounding aspects stated; (g) were the results gauged in a valid and trustworthy manner? and (h) was proper statistical analysis deployed. The 16 articles were compared to the questions,

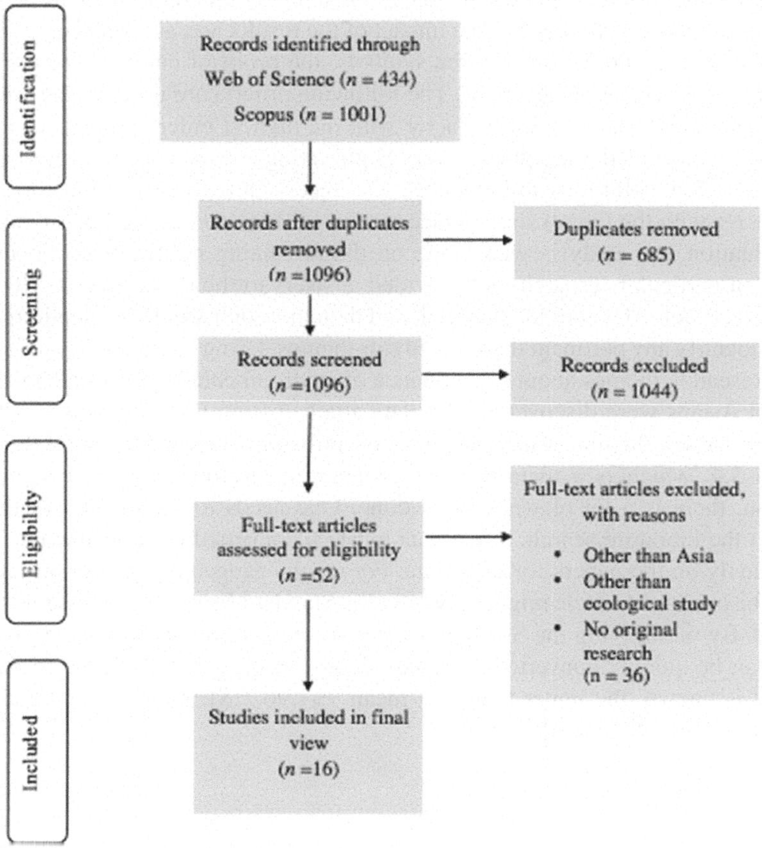

Fig. 5.1 Flow diagram of study selection for the review process

and yes/no responses were given. Table 5.2 provides information on the total risk using 'Low' and 'Moderate' indications. Every study's risk and quality were assessed independently by the authors (PPR and PM), and confusion was reduced by holding a consensus team meeting.

Using robvis (visualisation tools), the bias risk for the chosen studies was assessed. Every article's potential for bias risk was evaluated by two authors in separate analyses. Five questions made up the checklist: (a) bias resulting from the randomisation procedure; (b) bias resulting from deviations from intended interventions; (c) bias surfacing from the lack of outcome data; (d) bias resulting from the measurement of the outcome; and (e) bias surfacing from choosing the reported result. There were three levels of bias risk: "low," "high," and "moderate concerns" (Fig. 5.2a). The summary bar plot was made using the same dataset (Fig. 5.2b).

The Grading of Recommendations Assessment, Development, and Evaluation was employed. The quality of the evidence supporting every outcome was assessed using the 30 approaches. We examined the relationships for seven outcomes: minimum temperature, temperature, maximum temperature, rainfall, humidity, wind speed, and density of populace. The impact of the results was ascertained. Using the ⊕ AND/OR◯ combination of three symbols, the proof of certainty was rated as 'Moderate', 'High' or 'Very High'. The following criteria are used for the GRADE analysis's inputs: The correlation factor with the highest order significance among all the works included is assigned a 'Very High' rating. The parameterisation components that show solid measure evidence also receive a high point. The correlating perspective with the lowest significance receives 'Moderate' as the lowest measure.

Evaluation and analysis were done on the remaining publications. Focus was placed on particular research that provided answers to the raised queries. Reading the abstracts helped gather the material, and then thorough article (in-depth) reviews helped identify any pertinent themes and sub-themes. Using content analysis in qualitative research, themes about the influence of environmental aspects on COVID-19 cases in Asians were discovered. The data attained from the 16 papers was firstly recorded for conducting a meta-analysis by utilising Review Manager (RevMan) version 5.4, which is a meta-analysis programme. Before data are entered into RevMan, there is a lot of work to be done. One needs to spend a lot of time to conduct the literature search. Some data must be converted because meta-analyses rely heavily on the papers for their data. For results gauged using continuous variables, the published article might only offer the Standard Error or the 95% confidence interval. By multiplying the SE by the square root of the sample size, the Standard Error can be quickly converted to Standard Deviation. If just the 95% confidence interval is known, the upper bound $=$ mean $\pm$ 1.96 * SE method can be used to approximate the standard deviation.

Table 5.2 Risk bias assessment of the literature included in this study

Paper/ question	a. Were the criteria for inclusion in the sample clearly defined?	b. Were the study subjects and the setting described in detail?	c. Was the exposure measured in a valid and reliable way?	d. Were objective, standard criteria used for measurement of the condition?	e. Were confounding factors identified?	f. Were strategies to deal with confounding factors stated?	g. Were the outcomes measured in a valid and reliable way?	h. Was appropriate statistical analysis used?	Risk of Bias
Tosepu et al. (2020)	Yes	Yes	No	No	Yes	Yes	No	Yes	Low
Ahmadi et al. (2020)	Yes	Yes	Yes	Yes	Yes	Yes	Yes	Yes	Low
Şahin (2020)	Yes	Yes	Yes	Yes	Yes	Yes	Yes	Yes	Low
Coşkun et al. (2021)	Yes	No	Yes	No	Yes	Yes	Yes	Yes	Low
Ujiie et al. (2020)	Yes	Yes	No	No	No	No	No	Yes	Moderate
Haque and Rahman (2020)	Yes	Yes	No	No	No	No	Yes	Yes	Moderate
Ma et al. (2021)	Yes	Yes	Yes	Yes	Yes	Yes	Yes	Yes	Low
Qin et al. (2021)	Yes	Yes	Yes	Yes	Yes	Yes	Yes	Yes	Low

(continued)

Table 5.2 (continued)

Paper/ question	a. Were the criteria for inclusion in the sample clearly defined?	b. Were the study subjects and the setting described in detail?	c. Was the exposure measured in a valid and reliable way?	d. Were objective, standard criteria used for measurement of the condition?	e. Were confounding factors identified?	f. Were strategies to deal with confounding factors stated?	g. Were the outcomes measured in a valid and reliable way?	h. Was appropriate statistical analysis used?	Risk of Bias
Gupta et al. (2020)	Yes	Yes	Yes	Yes	Yes	Yes	Yes	Yes	Low
Meo et al. (2020)	Yes	Yes	Yes	Yes	No	No	Yes	Yes	Moderate
Rashed et al. (2020)	Yes	Yes	Yes	Yes	Yes	Yes	Yes	Yes	Low
Kodera et al. (2020)	Yes	Yes	Yes	Yes	Yes	Yes	Yes	Yes	Low
Sharma et al. (2020)	Yes	Yes	No	No	Yes	Yes	No	Yes	Low
Ladha et al. (2020)	Yes	Yes	No	No	Yes	Yes	Yes	Yes	Low
Sharif and Dey (2021)	Yes	Yes	Yes	Yes	No	No	Yes	Yes	Low
Pani et al. (2020)	Yes	Yes	Yes	Yes	Yes	Yes	Yes	Yes	Low

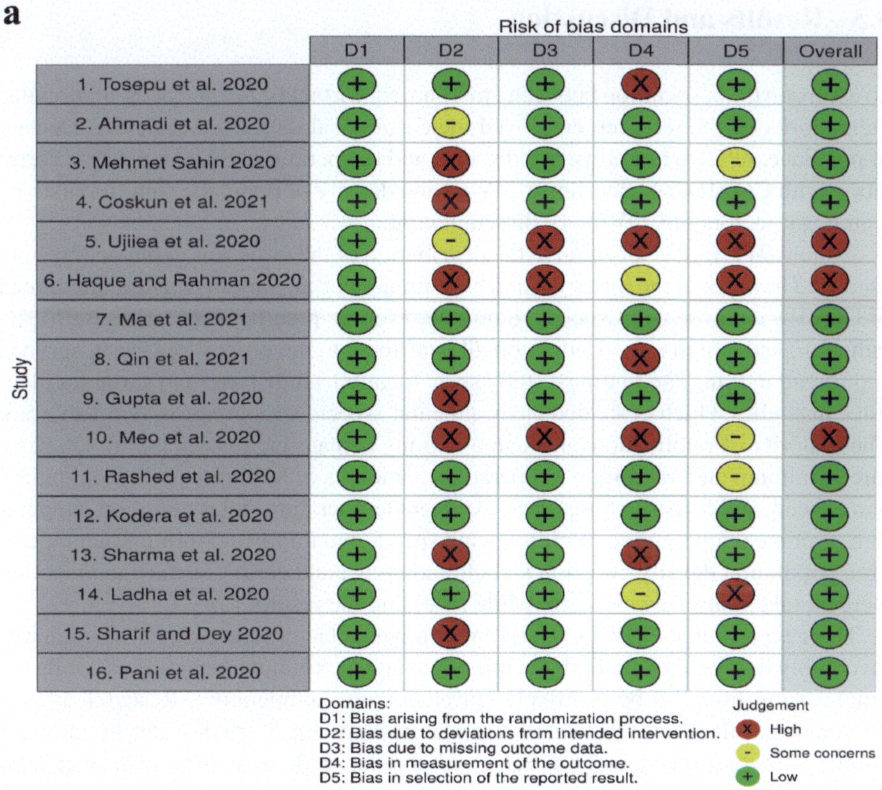

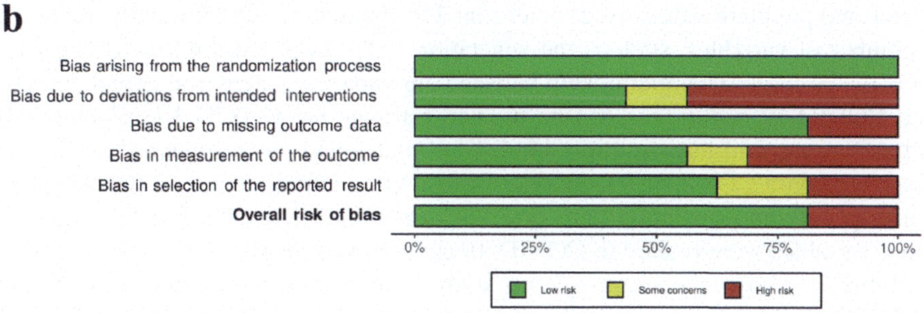

Fig. 5.2 Risk of bias based on **a** traffic light plot and **b** summary plot of ROB2 assessments created using robvis

5.3 Results and Discussion

To ascertain the association between environmental aspects, occurrences, and deaths, the majority of the research employed meteorological data from each day. Density of populace, relative humidity, wind speed, and temperature all had a positive correlation with COVID-19 occurrences. Wind and density of populace were found to be significant factors of COVID-19 dissemination.

Around 94% of the variation in the distribution of the virus was explained by both aspects. The onset of the monsoon and hot summer in Bangladesh can limit the spread of COVID-19. Low wind speeds, humidity, average precipitation, exposure to sun radiation, and a high rate of infection all contribute to the virus's long-term survival and spread in Iran. The findings, study span, location, study type, and variables identified in studies which evaluated environmental aspects are summarised in Table 5.3. The majority of publications evaluated various climate factors, including temperature, minimum and maximum temperatures, amounts of humidity and precipitation, wind speed, and density of populace. Ambient temperature, solar radiation temperature, average temperature, humidity, wind speed, maximum and minimum temperature, population density, location and climatic type, and earth's surface temperature were all taken into consideration while assessing the results.

There is proof that COVID-19 and environmental factors are related. One item was favourably connected with both the minimum and maximum temperatures, while ten articles were found to be positively correlated with temperature. Researchers also observed that there was a positive correlation between 5 articles and humidity, 4 articles and wind speed, and 5 articles and population density. A number of articles have found a negative link between precipitation, temperature, humidity, and wind speed. The factors employed for the study were humidity, temperature, rainfall, wind speed, and population density, as determined by the findings. Additionally, we noted a number of variables, such as the vegetation index, which did not associate with COVID-19 cases. The relationship between climate change, environmental aspects, and COVID-19 occurrence can be used to estimate the COVID-19 epidemic. To better comprehend how multiple kinds of climates and environment elements are associated with COVID-19 instances (for example, rainfall, temperature, humidity, wind speed, and population density), this systematic review studied how each of those variables were related to COVID-19 cases in various ways.

Table 5.4 shows the results of this study. Correlations among temperature and COVID-19 were gauged as (a) fixed-effect model: 0.31 (95% CI, 0.26–0.39) and (b) random effect model: 0.26 (95% CI − 0.12–0.55) with $p < 0.01$. Likewise, humidity and COVID-19 correlations were gauged as 0.64 (95% CI, − 0.55–0.75) and 0.58 (95% CI, − 0.36–0.95) for the fixed and random effect model; we noted that humidity had a positive association with COVID-19 at $p < 0.01$. Rainfall had a slight positive correlation with COVID-19 incidence: 0.41 (95% CI, 0.29–0.57) and 0.41 (95% CI, 0.30–0.57) for fixed and random, respectively. Wind speed was noted to be positively associated with the prevalence of COVID-19: 0.50 (95% CI, 0.32–0.77) and 0.51 (95% CI, 0.07–3.97). The correlation between density of populace

Table 5.3 Characteristics of selected articles

Paper	Year	Location	Samples and study period	Type of analysis	Variables	Main results
Tosepu et al.	2020	Jakarta, Indonesia	Jan–Mac 29, 2020	Spearman rank correlation	Min, max and average temperature, humidity and rainfall	Relationship between the average temperature and COVID-19 are significant
Ahmadi et al.	2020	Iran	Feb 19–March, 2020	The partial correlation coefficient (PCC)	Infected cases, population density, average temperature, rain, humidity, wind speed	Correlation between population density and COVID-19 are significant Areas with low humidity, wind speed, can support the viability of the virus, causing high infection rates
Şahin	2020	Turkey	March, 10–April 5, 2020	Spearman's correlation	Temperature, dew point, humidity, wind speed, population density	Temperature and wind speed with COVID-19 cases. Population density is prominent indicator

(continued)

Table 5.3 (continued)

Paper	Year	Location	Samples and study period	Type of analysis	Variables	Main results
Coşkun et al.	2021	Turkey	March 2020	Regression analysis, mediation analyses	Population density, rain, temperature, humidity, wind speed	The virus was effectively spread by population density and wind. Population density mediated the effect of wind speed on the number of COVID-19 cases
Ujiie et al.	2020	Japan	Jan 15–March 16, 2020	Poisson regression analysis	Population density, average temperature	Possible association between low temperature and increased risk of COVID-19 infection
Haque and Rahman	2020	Bangladesh	March 08–May 03, 2020	Linear regression	Temperature average, humidity	COVID-19 transmission is inhibited by high temperatures and relative humidity

(continued)

Table 5.3 (continued)

Paper	Year	Location	Samples and study period	Type of analysis	Variables	Main results
Ma et al	2021	Shanghai, China	Jan 21, 2020–February 29, 2020	Generalized additive model (GAM)	Temperature average, humidity, wind speed	Temperature humidity index (THI) and index of wind effect (K) showed significant negative relationships with the transmission of COVID-19
Qin et al.	2021	China	Jan 26–Feb 25, 2020	Longitudinal model, fixed-effect models	Average temperature, average humidity	High temperatures (-5 °C to 15 °C) and low relative humidity (<72%) may reduce the GR of COVID-19 outbreaks
Gupta et al.	2020	India	April 27th, 2020	Moment correlation, square regression, additive model	Air temperature, rainfall, actual humidity, wind speed, population density	Maximum transmission occurs with fewer wet condition. Provinces with the wet and very wet were less likely to be infected

(continued)

Table 5.3 (continued)

Paper	Year	Location	Samples and study period	Type of analysis	Variables	Main results
Meo et al.	2020	Gulf countries	Jan 29–May 15, 2020	Mean and standard error of mean (SEM), correlation coefficient, growth factor	Temperature, humidity	Negative correlation between cases with the increase humidity. But increasing the temperature, there was a correlation with cases
Rashed et al.	2020	Japan	Data 15 March–25 May 2020	Spearman's rank correlation, a correlation matrix with partial correlation. Linear regression	Spread and decay duration, population density, maximum temperature, absolute humidity	Population density influencing the distribution and damage pattern. The anticipated duration is derived through multivariate analysis and used to plan interventions in the event of a future pandemic

(continued)

Table 5.3 (continued)

Paper	Year	Location	Samples and study period	Type of analysis	Variables	Main results
Kodera et al.	2020	Japan	Data 22 February–25 May 2020	Spearman's rank correlation, correlation matrix with partial correlation probability	Population density, maximum temperature, humidity	As a consideration, population density should be factored. The percentage of air that is completely moist. Temperature, relative humidity, and number of cases per population were found to have a modest connection
Sharma et al.	2020	India	Data January 29, 2020–April 30, 2020	Pearson product moment correlation, spearman's rank correlation	Maximum and minimum temperature, specific humidity, temperature	Each variable had a significant relationship to the occurrence of COVID-19 cases
Ladha et al.	2020	Delhi, India	April 1–May 31, 2020	Linear regression	Max and average Temperature, average relative humidity	Maximum Temperature, average temperature and average relative humidity were did not significant with COVID-19 cases
Sharif and Dey	2021	Bangladesh	7 March 2020–14 August 2020	Spearman's rank correlation	Min, average and max. temperature, UV index, wind speed, rain fall, relative humidity	Temperature is the most significant factor related with COVID-19

and COVID-19 was gauged as 1.53 (95% CI, 1.34–1.75) for fixed and 1.71 (95% CI, 1.02–2.87) for the random model. Heterogeneity (I^2) was generally noted for temperature, humidity, wind speed and population density variables, i.e., 90%, 90%, 92% and 93%, respectively. Rainfall had total homogeneity, i.e. ($I^2 = 0$) concerning COVID-19 (zero x^2).

The results of our investigation showed a connection between COVID-19 incidence, fatality rates, and environmental aspects; a comparable rise in COVID-19 incidence during the summer suggests that the virus is not affected by temperature (Sahoo et al., 2021). The association between environmental variables and COVID-19 has never been fully understood, but this systematic review and meta-analysis is the first effort to do so. Ours is the first meta-analysis to assess how environmental aspects affect the incidence of COVID-19. We discovered an association between weather and COVID-19 based on the analysis. The most important climatic factors that have a significant impact on the occurrence of COVID-19 are humidity and temperature (Rashed et al. 2020; Sharma et al. 2020; Sharif and Dey 2021; Pani et al. 2020).

However, according to study from China (Wu et al. 2020) and Bangladesh (Haque and Rahman 2020), there was a negative correlation between temperature fluctuations and the rate of COVID-19 case rise. The findings concur with those of a study which suggested that human mobility served as a mediating factor between high temperatures and COVID-19 diffusion (Shao et al. 2021). The weather-centric methods to COVID-19 are used in all of the papers in this work. Research done between December 2019 and August 2020 is the basis for all papers.

Data on the influence of weather on coronavirus incidence and fatalities were gathered from more than 110 countries for our meta-analysis. This meta-analysis is much more useful in providing more detailed answers to correlation-related problems that have been extensively addressed in recent years because the articles conduct substantial research during the beginning and middle phases of COVID-19 in most countries. The proposed meta-analysis, which includes robvis, JBI, and GRADE

Table 5.4 Overall outcome

Outcome	Sample size	Pooled correlation (95% CI)		I^2(%)	x^2	*p* value
		Fixed effect model	Random effect model			
Temperature	718	0.31 [0.26, 0.39]	0.26 [0.12, 0.55]	90	0.96	< 0.01
Rainfall	127	0.41 [0.29, 0.57]	0.41 [0.30, 0.57]	0	0	0.99
Humidity	647	0.64 [0.55, 0.75]	0.58 [0.36, 0.95]	90	0.37	< 0.01
Wind speed	271	0.50 [0.32, 0.77]	0.51 [0.07, 3.97]	92	3.00	< 0.01
Population density	515	1.53 [1.34, 1.75]	1.71 [1.02, 2.87]	93	0.31	< 0.01

evidence profiles, adds significantly to the body of knowledge on the current COVID-19 incidence situation. We presumed that correlation values would be more effective in this meta-analysis because the relationship measurement approach is transparent. We used the fixed and random effect models as the foundation for our research, excluding the variability and dispersion of actual size impacts. Important factors affecting forest plots include temperature, humidity, wind speed, and people density. Additionally, there is a negative correlation between rainfall and COVID-19 incidence. Although we are unable to precisely remark on the aforementioned lack of significance, it is expected that increased rainfall improves relative humidity, leading to an increase in the importance of COVID-19 cases.

There are a number of drawbacks to our study, including the paucity of research on environmental aspects associated with COVID-19. Due to the limitations of our study, we had to conduct a meta-analysis on already published articles, some of which we found on different preprint servers. Thus, although we applied the robvis, JBI, and GRADE methodologies, the likelihood that those papers would be rejected was not properly calculated. Additionally, this technique can reveal risky behaviours and possible outbreak areas. Research on weather patterns, socio-environmental factors, and climate change can aid in our understanding of upcoming pandemics and disease outbreaks. We also suggest that hot regions with high relative humidity and average temperatures are more likely to have new COVID-19 outbreaks in the following days. Globally, the impending winter may offer some relief; nevertheless, additional research is required to fully validate the meta-analysis conclusions.

This study can help decision-makers comprehend how COVID-19 is spreading and how climatic factors are related to it. The study might, however, contain certain drawbacks. First, a large number of unknown asymptomatic positives in the populace could significantly change the results about the influence of the weather on COVID-19. Therefore, studies into the linkages between urbanisation and climate change are essential for stopping the rise in COVID-19 cases. Furthermore, limited daily tests may influence the quantity of confirmed new cases.

5.4 Conclusions

The majority of research on COVID-19 has been on COVID-19 in conjunction with climatic conditions, including humidity, temperature, wind speed, rainfall, and local populace. This well-known association can be used to determine the COVID-19 distribution. However, more knowledge about larger populations may have an impact on the climate, increasing the likelihood of COVID-19 occurrence and spread. Due to its connection to climate, wind speed should be included as a study variable as well. We noted that adverse weather increased the likelihood of COVID-19. The meta-analysis may enable governments and health organisations to use weather projections for making decisions before COVID-19 instances start increasing. This research area should be looked at by medical professionals and meteorologists since it might help to slow down the epidemic. It is important to examine COVID-19 instances to determine whether environmental variables such as humidity, temperature, rainfall, wind speed, and population density are associated with the frequency of cases.

Funding The Universiti Teknologi MARA Cawangan Selangor supported this work.

Conflict of Interest The author declare that they have no conflict of interest.

References

Baker RE, Yang W, Vecchi GA, Metcalf CJE, Grenfell BT (2020) Susceptible supply limits the role of climate in the early SARS-CoV-2 pandemic. Science 369:315–319

Coşkun H, Yildirim N, Gunduz S (2021) The spread of COVID-19 virus through population density and wind in Turkey cities. Sci Total Environ 751:141663

Dalziel BD, Kissler S, Gog JR, Viboud C, Bjornstad ON, Metcalf C, Grenfell BT (2018) Urbanization and humidity shape the intensity of influenza epidemics in U.S. cities. Science 362(6410):75–79

Gupta A, Banerjee S, Das S (2020) Significance of geographical factors to the COVID-19 outbreak in India. Modeling Earth Syst Environ 6(4):2645–2653

Haque SE, Rahman M (2020) Association between temperature, humidity, and COVID-19 outbreaks in Bangladesh. Environ Sci Policy 114:253–255

Huang C, Wang Y, Li X, Ren L, Zhao J, Hu Y, Zhang L, Fan G, Xu J, Gu X, Cheng Z, Yu T, Xia J, Wei Y, Wu W, Xie X, Yin W, Li H, Liu M, Xiao Y, Gao H, Guo L, Xie J, Wang G, Jiang R, Gao Z, Jin Q, Wang J, Cao B (2020) Clinical features of patients infected with 2019 novel coronavirus in Wuhan, China. Lancet 395:497–506

Kodera S, Rashed EA, Hirata A (2020) Correlation between COVID-19 morbidity and mortality rates in Japan and local population density, temperature, and absolute humidity. Int J Environ Res Public Health 17(15):5477

Ladha N, Bhardwaj P, Charan J, Mitra P, Goyal JP, Sharma P, Singh K, Misra S (2020) Association of environmental parameters with COVID-19 in Delhi, India. Ind J Clin Biochem IJCB 35(4):1–5

Liu J, Zhou J, Yao J, Zhang X, Li L, Xu X, He X, Wang B, Fu S, Niu T, Yan J, Shi Y, Ren X, Niu J, Zhu W, Li S, Luo B, Zhang K (2020) Impact of meteorological factors on the COVID-19 transmission: a multi-city study in China. Science Total Environ 726:138513

Ma Y, Cheng B, Shen J, Wang H, Feng F, Zhang Y, Jiao H (2021) Association between environmental factors and COVID-19 in Shanghai, China. Environ Sci Pollut Res Int 28(33):45087–45095

Meo SA, Abukhalaf AA, Alomar AA, Alsalame NM, Al-Khlaiwi T, Usmani AM (2020) Effect of temperature and humidity on the dynamics of daily new cases and deaths due to COVID-19 outbreak in gulf countries in Middle East region. Eur Rev Med Pharmacol Sci 24(13):7524–7533

Pani SK, Lin NH, RavindraBabu S (2020) Association of COVID-19 pandemic with meteorological parameters over Singapore. Sci Total Environ 740:140112

Prata DN, Rodrigues W, Bermejo PH (2020) Temperature significantly changes COVID-19 transmission in (sub)tropical cities of Brazil. Sci Total Environ 729:138862

Qi H, Xiao S, Shi R, Ward MP, Chen Y, Tu W, Su Q, Wang W, Wang X, Zhang Z (2020) COVID-19 transmission in Mainland China is associated with temperature and humidity: a time-series analysis. Sci Total Environ 728:138778

Qin L, Sun Q, Shao J, Chen Y, Zhang X, Li J, Chen M, Shia BC, Wu SY (2021) Association of temperature and relative humidity with the growth rate of the coronavirus disease 2019 epidemic. Am J Transl Res 13(6):5943–5955

Rabenau HF, Cinatl J, Morgenstern B, Bauer G, Preiser W, Doerr HW (2005) Stability and inactivation of SARS coronavirus. Med Microbiol Immunol 194:1–6

Rashed EA, Kodera S, Gomez-Tames J, Hirata A (2020) Influence of absolute humidity, temperature and population density on COVID-19 spread and decay durations: multi-prefecture study in Japan. Int J Environ Res Public Health 17:5354

Şahin M (2020) Impact of weather on COVID-19 pandemic in Turkey. Sci Total Environ 728:138810

Sahoo PK, Mangla S, Pathak AK, Salamao GN, Sarkar D (2021) Pre-to-post lockdown impact on air quality and the role of environmental factors in spreading the COVID-19 cases—a study from a worst-hit state of India. Int J Biometeorol 65(2):205–222

Sajadi MM, Habibzadeh P, Vintzileos A, Shokouhi S, Miralles-Wilhelm F, Amoroso A (2020) Temperature, humidity, and latitude analysis to estimate potential spread and seasonality of coronavirus disease 2019 (COVID-19). JAMA Netw Open 3(6):e2011834. https://doi.org/10.1001/jamanetworkopen.2020.11834

Shao W, Xie J, Zhu Y (2021) Mediation by human mobility of the association between temperature and COVID-19 transmission rate. Environ Res 194:110608

Sharif N, Dey SK (2021) Impact of population density and weather on COVID-19 pandemic and SARS-CoV-2 mutation frequency in Bangladesh. Epidemiol Infect 149:1–34

Sharma P, Singh AK, Agrawal B, Sharma A (2020) Correlation between weather and COVID-19 pandemic in India: An empirical investigation. J Public Affairs: e2222

Sehra ST, Salciccioli JD, Wiebe DJ, Fundin S, Baker JF (2020) Maximum daily temperature, precipitation, ultraviolet light, and rates of transmission of severe acute respiratory syndrome coronavirus 2 in the United States. Clin Infect Dis Official Publ Infectious Dis Soc Am 71(9):2482–2487

Tan J, Mu L, Huang J, Yu S, Chen B, Yin J (2005) An initial investigation of the association between the SARS outbreak and weather: with the view of the environmental temperature and its variation. J Epidemiol Community Health 59(3):186–192

Tosepu R, Gunawan J, Effendy DS, Ahmad LOAI, Lestari H, Bahar H, Asfian P (2020) Correlation between weather and covid-19 pandemic in Jakarta, Indonesia. Sci Total Environ 725:138436

Ujiie M, Tsuzuki S, Ohmagari N (2020) Effect of temperature on the infectivity of COVID-19. Int J Infect Dis IJID Official Publ Int Soc Infect Dis 95:301–303

Wang J, Tang K, Feng K, Lin X, Lv W, Chen K, Wang F (2021) Impact of temperature and relative humidity on the transmission of COVID-19: a modelling study in China and the United States. BMJ Open 11:e043863

Ward MP, Xiao S, Zhang Z (2020) The role of climate during the COVID-19 epidemic in New South Wales, Australia. Transboundary Emerging Dis 67(6):2313–2317

Wu Y, Jing W, Liu J, Ma Q, Yuan J, Wang Y, Du M, Liu M (2020) Effects of temperature and humidity on the daily new cases and new deaths of COVID-19 in 166 countries. Sci Total Environ 729:139051

Zhu N, Zhang D, Wang W, Li X, Yang B, Song J, Zhao X, Huang B, Shi W, Lu R, Niu P, Zhan F, Ma X, Wang D, Xu W, Wu G, Gao GF, Tan W (2020) A novel coronavirus from patients with pneumonia in China, 2019. N Engl J Med 382:727–733

Chapter 6
Covid-19 Pandemic: The Effect Towards Solid Waste Generation and Sustainable Environmental Approaches in Malaysia

Suhaila Mohd Omar and Ahmed Jalal Khan Chowdhury

Abstract Managing waste in a sustainable manner is a huge challenge for developing countries like Malaysia, with its rapidly expanding urban areas and growing population. To address this, the Malaysian government has been working hard to promote effective waste management methods that reduce landfill waste. However, the COVID-19 pandemic has brought a new set of environmental challenges, especially in terms of municipal and clinical waste generation and disposal. As a result of the pandemic, Malaysia has seen a reduction of 5% in municipal solid waste volume. However, there has been a worrying increase in food waste, which accounts for 36% of solid waste generated in 2021. In addition, the generation of clinical waste has increased significantly (43.9 %) and a large portion of medical waste comprises single-use plastic. Despite these challenges, Malaysia remains committed to environmental sustainability as evidenced by its established roadmap and strategic planning. The efficient management of solid waste is crucial in promoting better environmental health and requires active participation from communities and industries to support government policies and plans. This article reviews the impacts of the pandemic on municipal and clinical waste generation and disposal, as well as the ongoing sustainable effort in Malaysia.

Keywords COVID-19 pandemic · Landfill · Solid waste generation · Recycle · Environmental sustainability

---

S. M. Omar (✉)
Department of Biotechnology, Kulliyyah of Science, International Islamic University, Kuantan Campus, Jln. Sultan Ahmad Shah, 25200 Kuantan, Pahang, Malaysia
e-mail: osuhaila@iium.edu.my

Environmental Biology & Biotechnology Research Unit (EBBRU), Kulliyyah of Science, International Islamic University of Malaysia, Kuala Lumpur, Selangor, Malaysia

A. J. Khan Chowdhury
Faculty of Agriculture, Universiti Islam Sultan Sharif Ali (UNISSA), KM 33, Jalan Tutong, Kampung Sinaut, Tutong TB1741, Brunei Darussalam

D. Jones and A. J. Khan Chowdhury (eds.), *Environmental Resilience in the Pandemic Years 2020–2021*, Environmental Science and Engineering,
https://doi.org/10.1007/978-3-031-47757-7_6

6.1 Introduction

The COVID-19 pandemic has significantly changed our way of life, making it imperative to reassess the interactions between people and the environment and motivating us to make more sustainable decisions in our daily activities. The pandemic has altered the global socioeconomics landscape and Malaysia was not excluded. The government had acted by enforcing several restriction orders according to the level of infection across the nation, including a Movement Control Order (MCO), Conditional Movement Control Order (CMCO) and Recovery Movement Restriction Order (RMCO) as listed in Table 6.1.

Malaysians were exhorted to stay inside throughout the MCO period, and all forms of large-scale gatherings and movements were outright forbidden everywhere. All public and private educational facilities and locations were required to close, except for those engaged in providing essentials including utilities, healthcare, emergency services, food and other bare necessities, transportation, and logistics, as well as banking and finance. Travel restrictions were also put in place; neither tourists nor foreign nationals were allowed to enter Malaysia, nor were Malaysians allowed to travel outside. Malaysians coming back from abroad must go through a 14-day mandatory quarantine as well as health testing (Tay et al. 2021). The RMCO and later the National Recovery Plan was placed into effect as the infection cases began to decline, allowing cross-country travel and districts. On April 1, 2022, the Prime Minister announce the transition of the nation towards the Endemic Phase.

As of January 27, 2023, there were 752, 517,552 cumulative cases and 154, 596 new cases recorded worldwide (WHO 2023b). This number represents the Covid-19 statistic three years after the first COVID-19 case was reported in December 2019 close to Wuhan, China (WHO 2020). Severe Acute Respiratory Syndrome Coronavirus-2 (SARS-CoV-2), which affects the lower respiratory tract and behaves similarly to pneumonia, is the primary cause of this pandemic (Gorbalenya et al. 2020). Genetic mutations or viral recombination during genome replication cause changes to the genetic code of SARS-CoV-2, which is constantly evolving. While most variations may not affect the virus' properties, there is a possibility that new

Table 6.1 The list of restriction orders during the covid-19 pandemic in Malaysia

Timeline	Public health intervention due to covid-19 in Malaysia
18 March 2020–31 March 2020	Movement control order (MCO)
13 May 2020–9 June	Conditional movement control order (CMCO)
10 June 2020–31 March 2021	Recovery MCO (RMCO)
13 January 2021–31 May 2021	MCO by states
15 June 2021–31 December 2021)	National recovery plan (NRP)
1 April 2022	Transition to endemic phase of Covid-19

genetic alterations may impact the rate of disease spread, the severity of the accompanying sickness, the effectiveness of vaccines and therapeutic drugs, or other public health and social measures (WHO 2023a; Ministry of Health Malaysia 2022).

According to United Nations Environment Programme (2020), during the COVID-19 outbreak, more waste is being generated by hospitals, healthcare facilities and individuals than usual. The pandemic has also led to changes in waste management systems and procedures. There is also a significant increase in the amount of waste from single-use plastics (Peng et al. 2021). Although the number of Covid-19 cases had decreased after three years, some standard operating procedures to limit the spread of this disease have become a normal norm, including the use of face masks, hand sanitiser, and personal protective equipment for medical personnel. Meanwhile, socio-economic activities such as online business activities and online meetings also remained and gained popularity. This article reviews the impacts of the pandemic on the environment with a focus on municipal and clinical waste generation and disposal, as well as the ongoing sustainable effort in Malaysia.

6.2 Municipal Solid Waste (MSW) Management System in Malaysia

Rates of MSW generation are influenced by many factors including population growth, urbanisation, economic development, industrialization, living standards, societal customs, and regional climate (Mohd Omar et al. 2018). Malaysia was rated 27th out of 63 nations in the Institute for Management Development's (IMD) World Competitiveness Ranking 2020 report, which examined the state of the world's economy and its capacity to produce wealth (Malaysia Investment Development Authority 2020). Considered an upper-middle-income country in ASEAN, Malaysia has a population of 32.6 million in 2021. As of 2020, 75% of the population are living in urban area (Department of Statistics Malaysia 2022b) and this value is expected to increase as Asia's urban population was estimated to grow by more than 1.8 billion in 2050 (Asian Development Bank Institute 2020). Currently, per capita waste generation in Malaysia has doubled from 0.8 kg in 2005 to 1.17 kg in 2018 (Ministry of Housing and Local Government 2021). Municipal solid waste in Malaysia was collected from residential, commercial, institutional, commercial and city areas and the current waste generation is a challenge to the managing authorities.

In Malaysia, MSW is under the jurisdiction of the Ministry of Housing and Local Government (MHLG). Since September 2011, the country enforces the 672 Act: Solid Waste and Public Cleansing Management Act 2007 or also known as Act 672 emphasises waste separation at source and efficient solid waste treatment and disposal. To ensure the enforcement of Act 672, two agencies are established under the Solid Waste Management and Public Cleansing Corporation Act 2007 (Act 673) which are (1) National Solid Waste Management Department (Jabatan Pengurusan

Sisa Pepejal Negara—JPSPN) and (2) Solid Waste and Public Cleansing Management Corporation (SWCorp Malaysia). JPSPN is responsible for coordinating and cooperating with federal government agencies, state governments, local governments, associated private groups, and the general public to ensure effective solid waste management and public cleansing services. Meanwhile, SWCorp Malaysia complements the implementation of the related policies, plans and strategies (Moh and Abd Manaf 2017). The handling of MSW has been privatised to concession firms by the Federal Government of Malaysia by this act, taking over from state local authorities (LA). This action was taken to solve problems related to the lack of human and financial resources to control waste and an unstandardized MSW management method (Zainu and Songip 2021). All in all, the act was enacted to fortify the uniformity of law relating to the management of solid waste and public cleansing throughout Peninsular Malaysia and the Federal Territories of Kuala Lumpur, Putrajaya and Labuan (Act 672 2007; Moh and Abd Manaf 2017).

6.3 Clinical Waste Management in Malaysia

On the other hand, clinical waste is managed by the Ministry of Health (MOH), while hazardous waste is the responsibility of the Department of Environment (DOE). Inappropriate handling and management of clinical and biomedical waste raised substantial worries about the possibility of spreading disease and contaminating the environment. Clinical waste is waste that contains human or animal tissue, blood or body fluids, excretions, drugs, pharmaceutical products, soiled swabs or dressings, syringes, needles, and sharps—any waste that has come into contact or been mixed with any of the clinical types of waste and any waste unless rendered safe may prove hazardous to any person coming into contact with it (Department of Environment 2009). Malaysia's clinical waste is categorized into five main categories: clinical waste (CW); radioactive waste; chemical waste; pressurized containers; and general waste (Department of Environment 2009). Through Environmental Quality Act 1974, clinical and hazardous waste are listed as scheduled waste and categorized according to Environmental Quality (Scheduled Wastes) Regulations 2005 that have associated subsidiary legislation including: (a) The Environmental Quality (Prescribed Premises) (Scheduled Wastes Treatment and Disposal Facilities) Order 1989; and (b) The Environmental Quality (Prescribed Premises (Scheduled Wastes Treatment and Disposal Facilities) Regulations 1989 (Act 127 2006). Table 6.2, list out clinical waste type and designation in the said legislation.

Privatization of clinical waste management in Malaysia started back in 1996 when MOH hired three concessionaires: Radicare (M) Sdn. Bhd, Faber Medi-Serve Sdn. Bhd. and Pantai Medivest Sdn. Bhd., to oversee the medical waste management activities (Omar et al. 2018). After 15 years, the contract was renewed and upgraded. Currently, 5 licensed companies were responsible for handling CWM from hospitals, government clinics, or medical institutes including Radicare (M) Sdn Bhd, Faber Medi-Serve Sdn Bhd (now known as Edgenta Mediserve Sdn Bhd),

Table 6.2 Code and type of clinical waste stated in the environmental quality (scheduled wastes) regulations 2005

Waste code	Type of CW
SW403	Discarded drugs containing psychotropic substances or containing substances that are toxic, harmful, carcinogenic, mutagenic, or teratogenic
SW404	Pathogenic waste, CW, and quarantined materials
SW421	A mixture of scheduled wastes
SW422	A mixture of scheduled and non-scheduled wastes

Medivest Sdn Bhd, One Medicare Sdn Bhd dan Sedafiat Sdn Bhd. Additionally, another two companies, Kualiti Alam Sdn Bhd and Future NRG Sdn Bhd are the companies responsible for CWM from private hospitals and clinics (Agamuthu and Barasarathi 2021). The "cradle-to-grave" philosophy underpins the clinical waste management industry in Malaysia, where the private concessionaires take on most of the duties for each stage of the life cycle of medical waste including Healthcare Waste Management Service (HWMS), Linen and Laundry Service (LLS), Cleansing Services (CLS), Facility Engineering Maintenance Services (FEMS) and Biomedical Engineering Maintenance Services (BEMS). Collection, transport, disposal and supplying consumables. Healthcare professionals oversee the initial sorting and storing of the garbage (Department of Environment 2009).

The clinical waste treatment practices in Malaysia are mainly incineration and landfilling, while alternative methods such as autoclaving and recycling have not been widely embraced (Thien and Jusoh 2021). Incineration is typically effective for infectious and hazardous waste, but it carries the potential to release toxic gases into the atmosphere, putting human health at risk (Yong et al. 2019). While effective in treating most microorganisms autoclaving cannot be used to manage chemical and hazardous clinical waste such as chemotherapy waste, volatile or semi-volatile organic wastes, radioactive wastes, and mercury-containing waste (Thien and Jusoh 2021). Landfilling is the final option for clinical waste management but requires careful maintenance to avoid environmental pollution and health hazards. Recycling and recovery are sound environmental practices, reducing waste generation and the expenses associated with buying new equipment and disposal (Joseph et al. 2021). Microwave technology for clinical waste treatment poses a viable alternative to clinical waste management as it can save time, decrease costs, and have a minimal environmental impact. Evidence indicated that specially constructed microwave systems can sufficiently inactivate microorganisms; however, the process has to be strictly controlled by special microwave devices (Wang et al., 2020).

6.4 Food Waste Generation and Composition During the COVID-19 Pandemic

The COVID-19 outbreak had impacted waste management globally. The change in the waste composition and properties required careful handling especially when potential virus-contaminated substances were among the waste disposed of. As illustrated in Fig. 6.2, solid waste collected by the concessionaires in 2020 decreased by 5.3% compared to 2019. However, the amount had risen by 3.22% by 2021. In one of the studies on MSW generation before and after the implementation of MCOs in Peninsular Malaysia, a sharp decline from the end of March to the beginning of April 2020, was reported, indicating that the extensive MCO had affected waste generation due to less waste being produced in public areas (Brohan et al. 2021). However, with the removal of travel restrictions and the return of the majority of social and economic activities later, the volume of solid waste rises back (Brohan et al. 2021). The reduction of waste volume was in line with the impact of MCO's adoption towards economic growth, whereby 2020's Gross Domestic Product (GDP) recorded a growth that was negative by 5.6%. (2019: 4.3%) (Department of Statistics Malaysia 2021b). Figure 6.3 depicted the composition of solid waste in Malaysia from 2019 to 2021 (Ab Aziz et al. 2022).

Even before the COVID-19 pandemic, food waste constitutes the highest composition of solid waste in Malaysia ranging from 30.3 to 36% in 2019 and 2021 respectively (Ministry of Housing and Local Government 2019, 2020, 2021). Overall, Malaysian wasted approximately 17,000 tonnes of food every day (Fatimah 2021). It is hard to imagine this scenario when globally UNDP reported 821 million people estimated to be chronically undernourished in 2017 (UNDP 2023). There was a

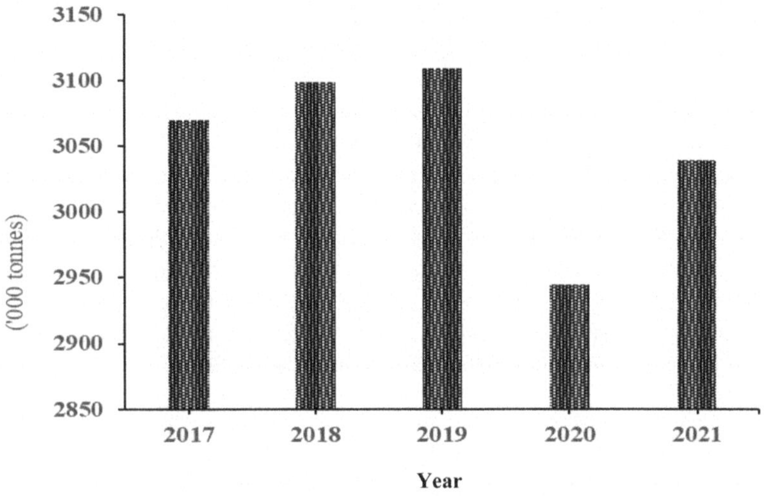

Fig. 6.2 Municipal solid waste collected by the concessionaires in Malaysia from 2017 to 2021

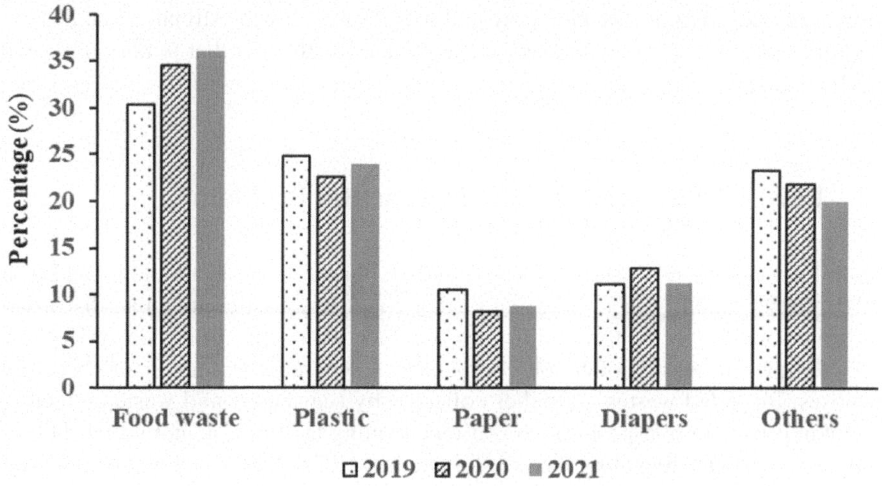

Fig. 6.3 Percentage and type of waste composition in Malaysia MSW from 2019 to 2021

decrease in food waste generated during the earlier MCO phase, attributed to most restaurants catering to takeaways, home deliveries, shorter operating hours, as well as the cancellation of all wedding functions and other events and gatherings. Buffet service, one of the major causes of food waste was not authorised during the MCO in hotels or restaurants (Bernama 2020). However, as movement restrictions were eased, and most economic sectors reopened food waste figures rose again. The data highlight an intriguing point since, despite a decrease in solid waste total volume in 2020 because of the lockdown, the percentage of food waste continued to rise. When food is lost or wasted, the resources that are being used for food processing, such as energy and water, are also wasted.

COVID-19-related policies including lockdowns and remaining at home to prevent the spread of the illness had increased internet usage in 2020 and 2021 by 5.4 and 7.2% respectively from 84.2% in 2019 (Department of Statistics Malaysia 2021a). Globally, between July 2019 and July 2020, there were approximately 376 million new users on social media, a growth of more than 10.5%, bringing the total number of users to 3.96 billion (Boshnjakoska 2022). The combination of increased social media usage and new marketing strategies of food manufacturers may result in more impulse buying of food products (Tariq et al. 2019). Besides, uncertainties about COVID-19 status have led to neurotic behaviours linked to unfavourable feelings including despair, anxiety, loneliness, dread of the unknown, and worry about product shortages (Yuen et al. 2020). These feelings have affected panicked impulse purchases and boosted social media usage (Gazali 2020). The increase in food waste in Malaysia during the COVID-19 pandemic was strongly influenced by social media usage, neuroticism, and impulse buying (Lahath et al. 2021). The fact that considerable amounts of food are produced but wasted by humans has substantial negative impacts on the environment, economy and social value. It was estimated that 8–10% of global greenhouse gas emissions are associated with food that is not consumed (UNEP 2021).

6.5 Plastic Waste and the Environmental Problem

While food and paper waste are biodegradable, plastic waste accumulates on Earth and decreases landfill space. Like many other countries, plastics are either discarded and transported directly to landfill sites together with other Malaysian MSW by approved private waste collection companies or separated at source for recycling purposes. Recycled wastes are either collected by scavengers and waste collectors or sent to private or non-profit organization recycling facilities by individuals before being transported to recycling factories via traders (JICA 2016; Moh and Abd Manaf 2017). Landfill known as a cheaper alternative is the most favoured form of solid waste management in Malaysia (Kaza et al. 2018). Approximately, Malaysian throw away 1.1 million tonnes of plastic every year. As illustrated in Fig. 6.2, plastics are the second largest type of waste in Malaysia MSW. The plastic industry in Malaysia is big and diversified with over 1300 plastic manufacturers producing goods ranging from automotive components to construction materials, housewares, and packaging products. In 2016 alone, resins of a value of 30 billion Malaysian ringgit (MYR) were exported to plastic producers around the world (World Bank Group 2021a).

The availability of single-use plastics promotes a convenient lifestyle. Plastic straws, single-use plastic bags, polystyrene plates and polypropylene utensils for takeaway food form the material basis of daily life, which is cheap, easily obtained and easy to dispose of (Heinrich Böll Foundation and Break Free from Plastic 2019). More than 75% of the material value of recyclable plastic is lost in Thailand, the Philippines and Malaysia, valued at USD 6 billion a year when single-use plastic is discarded rather than recovered and recycled (World Bank Group 2021a, b, c). Poor recovery and recycling of this material led to a huge amount of plastic waste polluting the environment, littering beaches and roadside, causing impact to the ecology, economy and aesthetics of terrestrial and aquatic ecosystems (Heinrich Böll Foundation and Break Free From Plastic 2019; Kwakwa 2021). The changes in consumption patterns and significant demand reductions for recycled products following low oil prices and economic slowdown during COVID-19 also impacted the plastic recycling industry (World Bank Group 2021a).

Before the outbreak of the COVID-19 pandemic, a lot of countries are developing a campaign to ban the usage of plastics and shifting towards paper packaging products instead of single-use plastics (Heinrich Böll Foundation and Break Free from Plastic 2019). However, the acceleration of the COVID-19 pandemic urged global efforts to increase PPE availability and raised production of plastics more than doubled (World Health Organization 2022). Single-use plastics provide enormous health benefits in terms of maintaining a sterile environment and, as a result, have become an integral part of our daily lives, particularly during the pandemic (Joseph et al. 2021). For example, in Malaysia, even though the composition of plastic waste collected in MSW (Fig. 6.3) was reduced in 2020 and 2021, during COVID-19, local plastic companies associated with the production of packaging materials for the food sector, or personal protective equipment (PPE), face shields, disposable medical aprons, plastic gloves, medical bottles, sprayers and hand sanitisers benefited from the surge in demand of these products (Teng 2020). Furthermore, the public's use of face masks and disposable gloves to prevent infection from others is widely observed together with improper disposal of waste in the environment harming birds and other wildlife as well as aquatic organisms (Agamuthu and Barasarathi 2021; Ammendolia et al. 2022). The use of plastics in syringes, intravenous bags, catheters, test kits, surgical tubing as well as in electronic device components has revolutionised the medical industries over the last century (Sastri 2014). Plastic has several advantages in the biomedical industry, including low cost, lightweight, ease of processing, and sterilizability (Joseph et al. 2021). However, the existing clinical waste guidelines indicated that plastic waste recycling from healthcare establishment is very limited and mostly end up in incinerator or landfill.

6.6 Clinical Waste and COVID-19

Each of approximately 296,496,809 global COVID-19 cases reported from March 2020 to January 2022 requires PPE usage by carers. Additionally, hundreds of millions more people have undergone COVID-19 testing due to exposure to COVID-19, travel, work, or leisure obligations (World Health Organization 2022). All these activities generated a massive amount of COVID-19-related waste which was partly infectious. Similarly in Malaysia, the increased use of disposable gloves, face masks, PPE by the medical staff and COVID-19 test equipment, as well as the opening of new health facilities, quarantine, vaccination centres, and laboratories had a direct impact on clinical waste management in 2021. Clinical waste increased by 43.9% in 2021 to 57.4 thousand tonnes, up from 39.9 thousand tonnes in 2020 (Table 6.3). Selangor (13.0 thousand tonnes), Sarawak (8.3 thousand metric tonnes), and W.P. Kuala Lumpur (5.9 thousand tonnes) produced the most clinical waste in the country (Department of Statistics Malaysia 2022a). These clinical wastes were generated on a daily basis from 144 government hospitals with 42,000 beds and 240 private hospitals with 16,000 beds (Agamuthu and Barasarathi 2021).

Understandably, the surge in clinical waste volume happened as many facilities and countries classified 100% of COVID-19 healthcare waste as hazardous, despite the 10–15% level typically generated from routine health service provision (World Health Organization 2022). Guidance was issued to treat and classified all waste generated by COVID-19 patients as infectious. This is despite the fact that SARS-CoV-2 is an enveloped virus, which means that it is inactivated relatively quickly by environmental factors such as sunlight or heat (Ratnesar-Shumate et al. 2020; World Health Organization 2022). While hospitals and other clinical institutions are following required standards and regulations for clinical waste management,

Table 6.3 Quantity of clinical waste by state, Malaysia, 2017–2021

State	2017 (tonnes)	2018 (tonnes)	2019 (tonnes)	2020 (tonnes)	2021 (tonnes)
Malaysia	**28,375.20**	**31,397.80**	**33,757.00**	**39,883.30**	**57,378.50**
Johor	2806.00	3097.50	3173.60	3989.60	5567.50
Kedah	1405.60	1466.30	1609.20	1659.70	2390.20
Kelantan	1162.30	1225.70	1355.50	1377.40	1857.40
Melaka	1195.80	1186.70	1895.00	1800.20	1594.00
Negeri Sembilan	1210.90	1242.90	1364.40	1645.70	3950.80
Pahang	1265.00	1315.90	1429.90	2999.40	2461.90
Perak	2193.50	2345.10	2644.20	792.8	4326.40
Perlis	245.4	251.3	262.9	33.1	1078.50
Pulau Pinang	2052.30	2344.70	2360.30	3336.00	2206.40
Sabah	1762.60	1849.30	1900.00	2606.00	3209.90
Sarawak	3086.90	3348.60	3683.80	4105.00	8265.60
Selangor	4370.70	6035.00	7343.20	9703.40	12,992.80
Terengganu	753	818.2	846.3	937.5	1287
W.P. Kuala Lumpur[1]	4808.10	4808.10	3822.40	4180.70	5892.90
W.P. Labuan	57.2	62.5	66.3	716.8	296.4

* Data from (Department of Statistics Malaysia 2022a)
The bold indicate the sum of all waste produce in all states, make up as the total of CW produced in Malaysia as a country

there are still deficiencies and weaknesses in their management that perhaps can be addressed to tackle the surge of clinical waste volume. These problems include a lack of instructions on waste segregation and practices, the intermingling of clinical waste with general waste, improper use of waste bags and containers, and low education and training among staff (Omar et al. 2018).

6.7 The Effort Towards Sustainability of Solid Waste Management in Malaysia

Malaysia has set ambitious targets for sustainability and the circular economy, aiming to achieve 40% recycling by 2025 and a 45% reduction in greenhouse gas emissions by 2030 (Ministry of Economy Malaysia 2021). In addition to Act 672, the Malaysian government has implemented numerous policies relating to waste management and circular economy (Fig. 6.4). Among the policies that highlight waste management and support the circular economy transition are the National Solid Waste

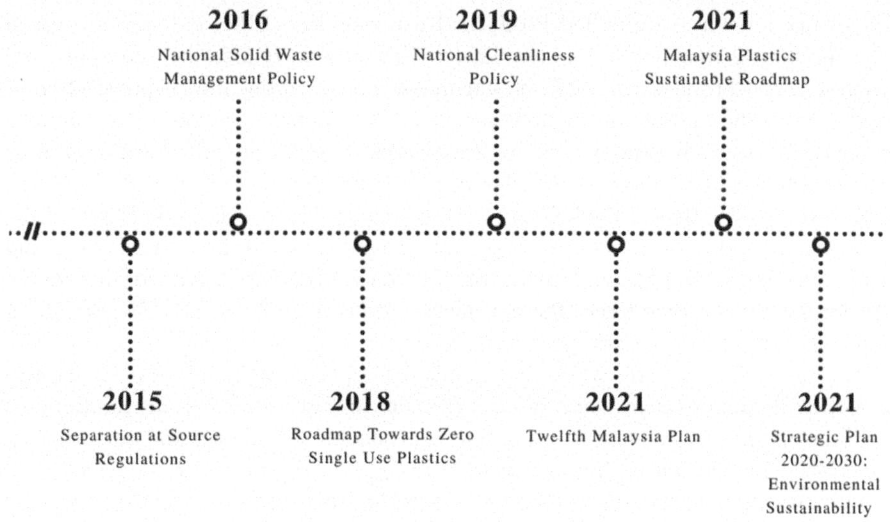

Fig. 6.4 Timeline of policies on waste management, circular economy and sustainability

Management Policy 2016, Green Technology Master Plan 2017–2030, Malaysia's Roadmap towards Zero Single-Use Plastics 2018–2030, National Cleanliness Policy: 2020–2030, Twelfth Malaysia Plan 2021–2025, and the Strategic Plan 2020–2030: Environmental Sustainability in Malaysia (BERNAMA 2021a; MAMPU 2021).

The Twelfth Malaysia Plan with the goal of "Keluarga Malaysia—Prosperous, Inclusive, Sustainable" which was launched in 2021, aligned the strategies and initiatives with the 2030 Agenda for Sustainable Development that was accepted by all the United Nations member states in 2015 and the effect of COVID-19 pandemic towards the nation. In focus area number 8: Advancing Green Growth; the government acknowledged the need to address the challenge of balancing socioeconomic development and environmental sustainability through the adoption of the circular economic model. The public and private sectors are encouraged to adopt and incorporate the Sustainable Development Goals (SDGs) and Environmental, Social and Governance (ESG) principles into their decision-making. The use of recycled materials and the recycling of production waste will be expanded, as will producers' responsibility for managing end-of-life products. All of these initiatives are expected to contribute to the growth of the green economy and the achievement of a low-carbon future (Ministry of Economy Malaysia 2021).

Malaysia is committed to address the issue of plastic waste pollution. This is evidenced by the establishment of a phased and holistic approach toward managing single-use plastic pollution by involving all stakeholders through its Roadmap Towards Zero Single-use Plastics 2018–2030. The action plan has three phases to tackle the 5 key problems in Malaysia: (i) severity of plastic pollution, (ii) low recycling rates of plastic waste, (iii) absence of environmentally friendly and (iv) cost-effective alternatives and the (v) absence of a uniform policy framework. A new framework known as Extended Producer Responsibility (EPR) is being discussed with the plastic manufacturers association with aims to reduce the economic and environmental burdens of waste management by extending the responsibility of producers to the end-of-life of their products. Additionally, as one of the actions towards Malaysia's Roadmap Towards Zero Single-Use Plastics which was approved by the cabinet in 2018, the Malaysia Plastics Sustainability Roadmap 2021–2030 was launched in 2021 (BERNAMA 2021b). At the regional level, Malaysia is part of the Coordinating Body of the Seas of East Asia (COBSEA) and the ASEAN Working Group on Coastal and Marine Environment (AWGCME), which are working on the development and protection of the marine environment and coastal areas, including addressing marine debris and plastic pollution.

The increase in plastic waste and demand for plastic markets serve as a good economic indicator for investor and government initiative to invest in technologies that converts plastic waste into value-added products such as fuel and construction materials. This will close the loop of the life cycle of plastic waste by achieving a sustainable circular economy (Khoo et al. 2021). To ensure the success of plastic recycling goals and the development of local solutions, the Government of Malaysia is targeting private sector participation and support. The success of waste recycling needs cooperation from various stakeholders including the government, manufacturers, and end-users. For example, was the success of a community recycling initiative by one of the food and beverage manufacturing companies. Their door-to-door collection has reached 112,000 households in Klang Valley and currently expanding to rural areas (The Star 2023). This is especially relevant in Malaysia since much of the recycling happens separately from the available solid waste management (SWM) (Khoo et al. 2021; World Bank Group 2021a).

6.8 Conclusions

The COVID-19 crisis emphasised the need to improve environmental health which requires a continuum of a comprehensive and integrated approach. Better environmental health plays an important role to reduce communities' vulnerability to pandemics and thus improve overall societal well-being and resilience (Suleimany et al. 2022). Apart from biodiversity conservation, and better air, water and sanitation quality, effective waste management is essential to minimise possible secondary impacts on health and the environment from COVID-19 (OECD 2020).

Proper segregation, pre-treatment and recycling of recyclable municipal solid waste and clinical waste will reduce the amount of waste sent to incineration and landfill abundantly. Nevertheless, post-COVID-19, as the economies bounced back to normal, the amount of waste that needs to be managed is already a challenge. Coupled with the waste-to-wealth aim, more green technology is being explored in Malaysia including waste-to-energy treatment plants (Yong et al. 2019). While the government has come out with many initiatives and policies, it's time for the industries and the nation to play their cards to ensure the sustainability of waste management in Malaysia. A proper coordination plan by the government and the industry to ensure the production of sustainable goods that can last longer, and are recyclable could be one of the tips to achieve sustainable development. Mindful consumption will be the best to replace the current 'throwaway' attitude in society. Malaysian should change the "take, make, and dispose" linear consumption pattern to a circular economy, where waste needs to be seen as a resource.

References

Ab Aziz NSA, Hasmady NII, Shafie FA, Mohd Yatim SR, Azmi A, Clark A (2022) Food waste and carbon footprint assessment of eateries in Kelantan, Malaysia. Malaysian J Med Health Sci 18(s15):1–8. https://doi.org/10.47836/mjmhs.18.s15.1

Agamuthu P, Barasarathi J (2021) Clinical waste management under COVID-19 scenario in Malaysia. Waste Manage Res J Sustain Circular Econ 39(1_suppl):18–26. https://doi.org/10.1177/0734242X20959701

Ammendolia J, Saturno J, Bond AL, O'Hanlon NJ, Masden EA, James NA, Jacobs S (2022) Tracking the impacts of COVID-19 pandemic-related debris on wildlife using digital platforms. Sci Total Environ 848:157614. https://doi.org/10.1016/j.scitotenv.2022.157614

Asian Development Bank Institute (2020) Solid waste management in developing Asia: prioritizing waste separation

BERNAMA (2020) As CMCO returns, so does food waste. Malay Mail. https://www.malaymail.com/news/malaysia/2020/10/20/as-cmco-returns-so-does-food-waste/1914411

BERNAMA (2021a) KASA launches environmental sustainability strategic plan. The New Straits Time. https://www.nst.com.my/news/nation/2021/06/696121/kasa-launches-environmental-sustainability-strategic-plan

BERNAMA (2021b) Launch of Malaysia plastics sustainability roadmap 2021–2030 is govt's effort to address waste pollution, says minister. Malay Mail. https://www.malaymail.com/news/malaysia/2021/12/10/launch-of-malaysia-plastics-sustainability-roadmap-2021-2030-is-govts-effor/2027521

Boshnjakoska R (2022) Latest social media marketing statistics in 2022. https://review42.com/resources/social-media-marketing-statistics/#

Brohan MA, Dom NC, Ishak AR, Abdullah S, Salim H, Ismail SNS, Precha N (2021) An analysis on the effect of coronavirus (COVID-19) pandemic movement control order (MCOS) on the solid waste generation in Peninsular Malaysia. Environ Sci Pollut Res 28(46):66501–66509. https://doi.org/10.1007/s11356-021-17049-6

Department of Environment (2009) Guideline on the handling and management of clinical wastes in Malaysia, 3rd edn. Government of Malaysia

Department of Statistics Malaysia (2021a) ICT use and access by individuals and households survey report, Malaysia, 2021. https://www.dosm.gov.my/v1/index.php?r=column/cthemeByCat&cat=395&bul_id=bCs4UlNSQktybTR3THZ3a0RzV2RkUT09&menu_id=amVoWU54UTl0a21NWmdhMjFMMWcyZz09

Department of Statistics Malaysia (2021b) Malaysia economic performance fourth quarter 2020

Department of Statistics Malaysia (2022a) Compendium of environment statistics, Malaysia, 2022. https://www.dosm.gov.my/

Department of Statistics Malaysia (2022b) Current population estimates, Malaysia, 2021. https://newss.statistics.gov.my/

Environmental Quality Act 1974 (2006) (testimony of Act 127)

Fatimah Z (2021). Daily food waste staggering. The Star. https://www.thestar.com.my/news/nation/2021/05/20/daily-food-waste-staggering

Gazali HM (2020) The covid-19 pandemic: factors triggering panic buying behaviour among consumers in Malaysia. Labuan Bull Int Bus Fin (LBIBF) 18(1):84–95. https://doi.org/10.51200/lbibf.v18i1.2690

Gorbalenya AE, Baker SC, Baric RS, de Groot RJ, Drosten C, Gulyaeva AA, Haagmans BL, Lauber C, Leontovich AM, Neuman BW, Penzar D, Perlman S, Poon LLM, Samborskiy D, Sidorov IA, Sola I, Ziebuhr J (2020) Severe acute respiratory syndrome-related coronavirus: the species and its viruses—a statement of the coronavirus study group. BioRxiv, 2020.02.07.937862. https://doi.org/10.1101/2020.02.07.937862

Heinrich Böll Foundation and Break Free From Plastic (2019) Plastic atlas: facts and figures about the world of synthetic polmers. https://www.boell.de/sites/default/files/2020-01/Plastic%20Atlas%202019%202nd%20Edition.pdf

Japan International Cooperation Agency (JICA) (2016) The study on national waste minimisation in Malaysia: final report summary. https://openjicareport.jica.go.jp/pdf/11831245_01.pdf

Joseph B, James J, Kalarikkal N, Thomas S (2021) Recycling of medical plastics. Adv Industr Eng Polymer Res 4(3):199–208. https://doi.org/10.1016/j.aiepr.2021.06.003

Kaza S, Yao LC, Bhada-Tata P, Van Woerden F (2018) What a waste 2.0 : a global snapshot of solid waste management to 2050. http://hdl.handle.net/10986/30317

Khoo KS, Ho LY, Lim HR, Leong HY, Chew KW (2021) Plastic waste associated with the COVID-19 pandemic: crisis or opportunity? J Hazard Mater 417:126108. https://doi.org/10.1016/j.jhazmat.2021.126108

Kwakwa V (2021) Plastic waste is a growing menace, and a wasted opportunity. https://blogs.worldbank.org/eastasiapacific/plastic-waste-growing-menace-and-wasted-opportunity

Lahath A, Omar NA, Ali MH, Tseng M-L, Yazid Z (2021) Exploring food waste during the COVID-19 pandemic among Malaysian consumers: the effect of social media, neuroticism, and impulse buying on food waste. Sustain Prod Consump 28:519–531. https://doi.org/10.1016/j.spc.2021.06.008

Malaysia Investment Development Authority (2020) Malaysia 27th most competitive nation. https://www.mida.gov.my/mida-news/malaysia-27th-most-competitive-nation/

Ministry of Economy Malaysia (2021) Speech by the prime minister in the Dewan Rakyat: twelfth Malaysia plan 2021–2025. https://rmke12.epu.gov.my/en/documents/speech

Ministry of Health Malaysia (2022) Gelombang varian Omicron. https://covid-19.moh.gov.my/semasa-kkm/2022/02/gelombang-varian-omicron

Ministry of Housing and Local Government Malaysia (MHLG) (2019) KPKT statistics 2019. https://www.kpkt.gov.my/index.php/pages/view/700?mid=586

Ministry of Housing and Local Government Malaysia (MHLG) (2020) KPKT statistics 2020. https://www.kpkt.gov.my/index.php/pages/view/700?mid=586

Ministry of Housing and Local Government (2021) KPKT statistic 2021

Moh YC, Abd Manaf L (2017) Solid waste management transformation and future challenges of source separation and recycling practice in Malaysia. Resour Conserv Recycl 116:1–14. https://doi.org/10.1016/j.resconrec.2016.09.012

Mohd Omar S, Chowdhury AJK, Abdulahi Hashi A (2018) Islamic ethics of waste management towards sustainable environmental health. IIUM Med J Malaysia 17(1). https://doi.org/10.31436/imjm.v17i1.1024

OECD (2020) Environmental health and strengthening resilience to pandemics. OECD Publishing. https://www.oecd.org/coronavirus/policy-responses/environmental-health-and-strengthening-resilience-to-pandemics-73784e04/

Omar D, Nazli SN, Karuppannan S (2018) Clinical waste management in Malaysia. J ASIAN Behavioural Stud 3(7):11. https://doi.org/10.21834/jabs.v3i7.253

Peng Y, Wu P, Schartup AT, Zhang Y (2021) Plastic waste release caused by COVID-19 and its fate in the global ocean. Proc Nat Acad Sci 118(47). https://doi.org/10.1073/pnas.2111530118

Ratnesar-Shumate S, Williams G, Green B, Krause M, Holland B, Wood S, Bohannon J, Boydston J, Freeburger D, Hooper I, Beck K, Yeager J, Altamura LA, Biryukov J, Yolitz J, Schuit M,

Wahl V, Hevey M, Dabisch P (2020) Simulated sunlight rapidly inactivates SARS-CoV-2 on surfaces. J Infect Dis 222(2):214–222. https://doi.org/10.1093/infdis/jiaa274
Sastri VR (2014) Plastics in medical devices, 2nd edn. Elsevier. https://doi.org/10.1016/B978-1-4557-3201-2.00001-X
Solid Waste and Public Cleansing Management Act 2007 (Act 672) (2007) (testimony of Act 672)
Suleimany M, Mokhtarzadeh S, Sharifi A (2022) Community resilience to pandemics: an assessment framework developed based on the review of COVID-19 literature. Int J Disaster Risk Reduction 80:103248. https://doi.org/10.1016/j.ijdrr.2022.103248
Tay YL, Abdullah Z, Chelladorai K, Low LL, Tong SF (2021) Perception of the movement control order during the COVID-19 pandemic: a qualitative study in Malaysia. Int J Environ Res Pub Health 18(16):8778. https://doi.org/10.3390/ijerph18168778
The Star (2023) Expanding recycling to rural areas. The Star. https://www.thestar.com.my/metro/metro-news/2023/02/03/expanding-recycling-to-rural-areas
Tariq A, Wang C, Tanveer Y, Akram U, Bilal M (2019) Online impulse buying of organic food: a moderated (website personality) mediation (social appeal) process. Int J Inf Syst Change Manage 11(1):3. https://doi.org/10.1504/IJISCM.2019.101646
Teng LJ (2020) Not all players benefiting from pandemic. The Edge Markets
The Malaysian Administrative Modernisation and Management Planning Unit (MAMPU) (2021) *Government policies*. https://www.malaysia.gov.my/portal/category/298
Thien CY, Jusoh MNH (2021) Overview of clinical waste management in Malaysia. Front Water Environ 1(1)
UNDP (2023) Goal 2: zero hunger. https://www.undp.org/sustainable-development-goals/zero-hunger
United Nation Environment Program (UNEP) (2021) UNEP food waste index report. https://www.unep.org/resources/report/unep-food-waste-index-report-2021
United Nations Environment Programme (2020) Waste management during the COVID-19 pandemic: from response to recovery
Wang J, Shen J, Ye D, Yan X, Zhang Y, Yang W, Li X, Wang J, Zhang L, Pan L (2020) Disinfection technology of hospital wastes and wastewater: suggestions for disinfection strategy during coronavirus Disease 2019 (COVID-19) pandemic in China. In: Environmental pollution, vol 262. Elsevier Ltd. https://doi.org/10.1016/j.envpol.2020.114665
WHO (2020, January 5) Pneumonia of unknown cause – China. https://www.who.int/emergencies/disease-outbreak-news/item/2020-DON229
WHO (2023a) Tracking SARS-CoV-2 variants. https://www.who.int/activities/tracking-SARS-CoV-2-variants
WHO (2023b) World Health Organisation WHO coronavirus disease (COVID-19) dashboard. https://covid19.who.int/
World Bank Group (2021a) Market study for Malaysia: plastics circularity opportunities and barriers. https://openknowledge.worldbank.org/handle/10986/35296
World Bank Group (2021b) Market study for Philippines: plastics circularity opportunities and barriers. http://hdl.handle.net/10986/35295
World Bank Group (2021c) Market study for Thailand: plastics circularity opportunities and barriers. http://hdl.handle.net/10986/35114
World Health Organization (2022) Global analysis of healthcare waste in the context of COVID-19: status, impacts and recommendation. https://www.who.int/publications/i/item/9789240039612
Yong ZJ, Bashir MJ, Ng CA, Sethupathi S, Lim JW, Show PL (2019) Sustainable waste-to-energy development in Malaysia: appraisal of environmental, financial, and public issues related with energy recovery from municipal solid waste. Processes 7(10):676.https://doi.org/10.3390/pr7100676
Yuen KF, Wang X, Ma F, Li KX (2020) The psychological causes of panic buying following a health crisis. Int J Environ Res Public Health 17(10):3513. https://doi.org/10.3390/ijerph17103513
Zainu ZA, Songip AR (2021) Policies, challenges and strategies for municipal waste management in Malaysia. J Sci Technol Innov Policy 3(1):10–14. https://doi.org/10.11113/jostip.v3n1.18

Chapter 7
Environmental Benefits Evaluation of Construction CO_2 Emission Reduction Management Using System Dynamics Approach

Haixin Wang, Shengsong Wei, Xin Chen, Zuhe Wang, and Qian Xu

Abstract With the rapid development of urbanization, a huge amount of greenhouse-gas has been generated from increasing higher amounts of construction activities, which has significant negative impacts on the environment if they are not properly managed. Therefore, more and more scholars start to research the CO_2 emissions reduction of construction industry. However, it has been determined that existing studies with respect to CO_2 emissions reduction are mainly carried out from a static perspective, without considering the dynamic and interdependent nature of the whole CO_2 emissions reduction system. Therefore, based on literature review and interview survey, Vensim software was used to establish a system dynamic model of construction CO_2 emissions reduction management at the construction stage and simulate the environmental benefits of construction CO_2 emissions reduction management. The application of the proposed model is shown by using a case. The simulation results indicate that the CO_2 emissions reduction management can reduce 10.84% of CO_2 emissions. In the meantime, the CO_2 emissions reduction management achieves good environmental benefits including saving energy consumption of 59,718.5 kgce, saving waste landfill of 8434.28 m^3 and the reduction of global warming potential of 3620.51 $tCO_{2\text{-}e}$. Scenario analysis reveals that using low-carbon materials has the optimal reduction effect in single measure, which can reduce 16.50% of CO_2emissions. Compared with a single measure, the maximum emissions reduction of multiple measures can reach 21.05%. Therefore, in order to realize best environmental benefits of construction CO_2 emissions reduction management, multiple measures should be used simultaneously, such as using low-carbon materials, recycling and reusing materials and improving operation efficiency

H. Wang · S. Wei (✉) · X. Chen · Z. Wang
College of Civil Engineering and Architecture, Shandong University of Science and Technology, Qingdao 266510, China
e-mail: wss17863657557@163.com

Q. Xu
School of Management Engineering, Shandong Jianzhu University, Jinan 250101, China

D. Jones and A. J. Khan Chowdhury (eds.), *Environmental Resilience in the Pandemic Years 2020–2021*, Environmental Science and Engineering,
https://doi.org/10.1007/978-3-031-47757-7_7

of construction equipment. This study can help construction professionals determine the optimal emissions reduction strategy and provide references for project managers in assessing the reduction outcomes of construction projects and the environmental benefits.

Keywords CO_2 emissions reduction management · Environmental benefits · System dynamic

7.1 Introduction

Global warming has become a significant environmental problem and it is considered as one of the biggest challenges faced by the society (Lai et al. 2012). CO_2 has been recognized as the main contributor for global warming (Ozkaymak et al. 2017). With increasing global urbanization, construction activities have intensified and construction industry is now the primary emitter of carbon (Guo et al. 2017). The construction industry consumes 40% of the world's energy each year and produces about 33% of the world's greenhouse gases (Sbci 2009). The United Nations Environment Programme (UNEP) has declared that due to the rapid urbanization and the inefficiency of existing building stock, greenhouse gas emissions will be duplicated by 2030 unless the measures of emissions reduction are implemented (UN Environment Programme 2016). Therefore, the development and use of adequate reduction strategies will play an important role in construction greenhouse gas emissions reduction.

Most relevant studies in this domain focused on CO_2 emissions or reduction measures during the entire life cycle of buildings or several individual stages of a life cycle. For example, Basbagill et al. applied life cycle assessment (LCA) to early decision-making in order to inform designers of the relative environment impact important of building component material and dimensioning choices and emphasized that early decision-making can significantly reduce embodied carbon emissions (Basbagill et al. 2013). Others studies have focused on the adoption of advanced tools and architectural structures at the design stage, such as building information model (BIM) (Inyim et al. 2015; Clevenger and Khan 2014) and different structural building (Teng et al. 2018; Zhang and Zheng 2020). Rasmussen et al. researched the impact of different design strategies on the lower embodied carbon and proposed to integrate various design aspects of entire life cycle of buildings to comprehensively optimize building design (Rasmussen et al. 2020). There were also some studies about CO_2 emissions and reduction measures at the operation stage. Cheng et al. pointed out that approximately 80% of energy use and greenhouse gas emissions are generated during the operation stage of building (Cheng et al. 2008), whereas only 10–20% are from material manufacturing, construction, and demolition. The emissions reduction measures at the operation stage were mainly the development of energy conservation technologies (Zhang et al. 2011; Zhou et al. 2016) and improvement of polices and operation management (Ahn et al. 2010; Papachristos et al. 2020). However, due to existing energy saving codes or other polices, the CO_2 emissions that occur during the operation has declined dramatically over time and thus the relative contribution of construction stage emissions and impacts becomes more important. Guggemos and Horvath also emphasized that future studied should concentrate on the construction stage and should not only research the design stage or operation stage (Guggemos and Horvath 2006). Therefore, CO_2 emissions reduction measures and environmental impacts at the construction stage must be analyzed.

The construction stage should not be neglected because of its negative environmental impacts with excessive consumption of resources and huge amount of construction waste (Ortiz et al. 2010). Reduction measures at the construction stage is divided into two aspects: (1) they concentrated on the plan selection of building materials to reduce CO_2 emissions, and (2) quantitative analysis of the impact of construction equipment operating efficiency on CO_2 emissions reduction potential by field studies. For example, CO_2 emissions can be effectively reduced by as much as 30% at the construction stage by selection of low-carbon materials and can save raw materials and energy consumption (González and Navarro 2006; Sadok et al. 2022; Naqi and Jang 2019). Researchers found that converting already made products consumes fewer resources and produces fewer CO_2 emissions than obtaining raw materials from scratch and producing a totally new product (Chau et al. 2012; Saghafi and Teshnizi 2011; Ng and Chau 2015). Fuel can be saved and CO_2 emissions can be reduced by eliminating idle time and improving construction equipment efficiency (Lewis et al. 2012; Szamocki et al. 2019). Szamocki et al. also pointed out that appropriate storage area can reduce huge amount of CO_2 emissions (Szamocki et al. 2019). Amor et al. found that local sourcing materials can significantly influence the environmental performance of embodied carbon emissions (Padilla-Rivera et al. 2018).

In construction projects, CO_2 emissions reduction management is a complex process and requires an assessment of many complex interactions such as design planning, materials production, transportation systems, site construction, operation and maintenance and interdependencies in the system. Therefore, to improve the scientific understanding of integrated systems, a simulation model is needed that explicitly considers the dynamic and complexity of CO_2 emissions reduction management. System Dynamic (SD) approach offers possibility to address system complexity problems. For example, Bank et al. (2010) tried to combine BIM and SD to provide decision-making for the sustainability of building during the design and operation stage. However, previous studies only focused on the design and operation stage and did not systematically consider and analyze the CO_2 emissions reduction management at the construction stage, and research dynamic interaction between different factors, and thus the effect of emissions reduction cannot be fully revealed.

To fulfill this knowledge gap, the main aim of this study is to simulate the effects of management strategies on CO_2 emissions reduction and environment with the aid of system dynamic approach. In this paper, the key factors of CO_2 emissions reduction management at the construction stage were identified by literature review and interview survey, an environmental benefit simulation of CO_2 emissions reduction management at the construction stage is developed to simulate how the dynamic relations among the factors and their variations influence the CO_2 emissions management and the environmental benefits.

7.2 System Dynamic and Application in Construction CO_2 Emissions Management Studies

SD is an interdisciplinary and comprehensive subject that studies information feedback, recognizes system problems and solves system problems. It deals with complex system problems with qualitative and quantitative methods (Wang 2009). A SD model is to simulate a real world system by extracting the real system structure and the key variables and it is capable for observing the dynamic behaviors of systems and interactions among system elements from a holistic perspective (Yuan and Wang 2014).

In recent years, SD has been extensively applied across a wide range of domain and researchers have applied it to some issues related to sustainable development. For example, Lim et al. (2019) developed a SD model of life cycle cost and CO_2 emission and systematically analyzed cost-effective emissions reduction measures. Thompson and Bank (2010) applied SD model as a decision-making tool for building design and operation.

7.3 Model Development

The CO_2 emissions reduction management at the construction stage is a complex dynamic system involving various reduction measures, environmental benefits and their interdependence. Base on literature (Ma 2017; Sandanayake et al. 2017), this model is divided into two subsystems: (1) CO_2 emissions reduction management subsystem, (2) environmental benefits subsystem. In this model, the variables of CO_2 emissions reduction management subsystem can cause variations of CO_2 emissions reduction and its effects are directly reflected in the environmental benefits. On the contrary, the results of environmental benefits assessment of CO_2 emissions reduction will also affect CO_2 emissions reduction management subsystem by a feedback loop.

7.3.1 CO_2 Emissions Reduction Management Subsystem

Generally, the structure of a SD model is described by causal-loop diagrams which capture the major feedback mechanisms. Causal-loop diagram is a conceptual tool

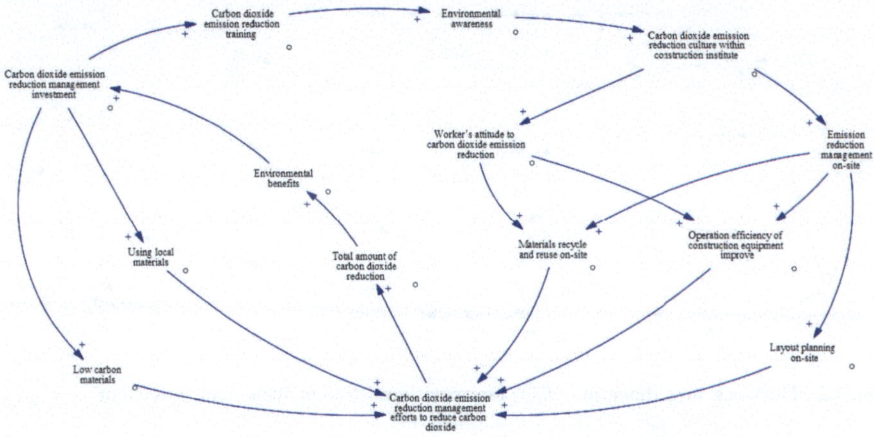

Fig. 7.1 The causal-loop diagram of CO_2 emissions reduction management subsystem

which reveals a dynamic process and can show how the system is dynamically influenced by the interactions of all variables. Figure 7.1 shows causal-loop diagram of CO_2 emissions reduction subsystem which includes seven positive feedback loops and reveals the feedback mechanisms inherent in the CO_2 emissions reduction subsystem.

In positive feedback loops, a change in any variables within causal-loop can directly affect itself in a positive way. For example, in the feedback loop R1 (CO_2 emissions reduction management efforts to reduce CO_2 emissions → Total amount of CO_2 emissions reduction → Environmental benefits → CO_2 emissions reduction management investment → Low-carbon materials → CO_2 emissions reduction management efforts to reduce CO_2 emissions), as various emissions reduction measures are taken, the effect of CO_2 emissions reduction management will be reflected in the total amount of CO_2 emissions reduction. The improvement of the total of emissions reduction will bring good environmental benefits which and then increase the investment of emissions reduction measures. The same logic is applied to feedback loops, R2, R3, R4, R5, R6, R7.

Based on causal-loop diagram of CO_2 emissions reduction management subsystem, all key variables that affect CO_2 emissions reduction management are identified. Then, Vensim software is used to convert causal-loop diagram into a stock-flow diagram. The stock-flow diagram of CO_2 emissions reduction is shown in Fig. 7.2. All descriptions of the variables are listed in Appendix 1.

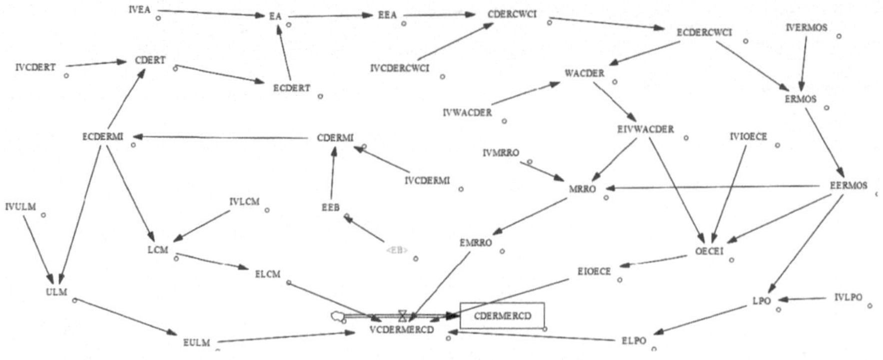

Fig. 7.2 The stock-flow diagram of CO_2 emissions reduction management subsystem

7.3.2 *Environmental Benefits Subsystem*

There are many types of environmental impacts caused by construction process, such as materials and energy consumption, waste production and global warming. Based on literature (Ding et al. 2018; Cao et al. 2015), this paper selects four evaluation index of environmental benefits which include (1) saving of materials consumption, (2) land space saving from landfilling reduction, (3) saving energy, and (4) reduction of global warming potential.

The causal-loop diagram of environmental benefits subsystem is shown in Fig. 7.3. Eight causal-loop diagrams are presented and analyzed and all causal-loop diagrams are positive feedback loops. In positive feedback loop R1 (CO_2 emissions reduction management efforts to reduce CO_2 emissions → Total amount of CO_2 emissions reduction → The increase of materials recycle and reuse on-site → Land space saving from landfilling reduction → Environmental benefits → CO_2 emissions reduction management efforts to reduce CO_2 emissions), as various emissions reduction measures are taken, the effect of CO_2 emissions reduction management will be reflected in the total amount of CO_2 emissions reduction. The increasing of total amount of emissions reduction will not only increase amounts of materials recycle and reuse, but also reduce the amount of waste which is transported to landfill and save landfilling space.

Based on the casual loop diagram of environmental benefits subsystem, the stock-flow diagram of environmental benefits subsystem is shown in Fig. 7.4. All descriptions of the variables are listed in Appendix 2.

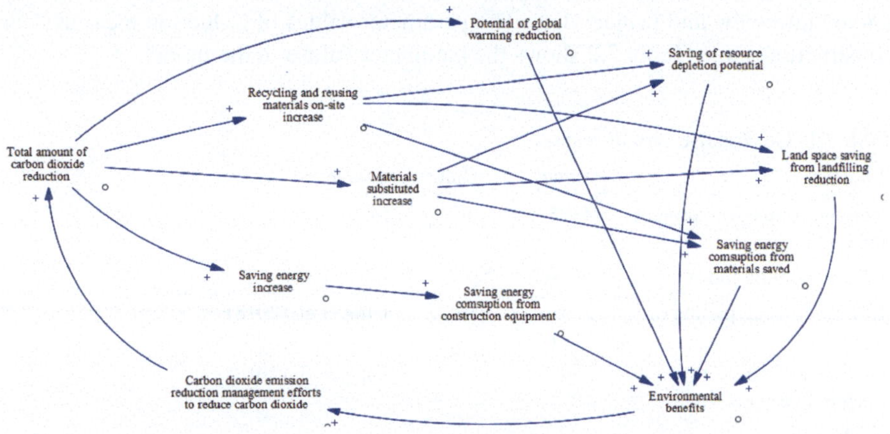

Fig. 7.3 The causal-loop diagram of environmental benefits subsystem

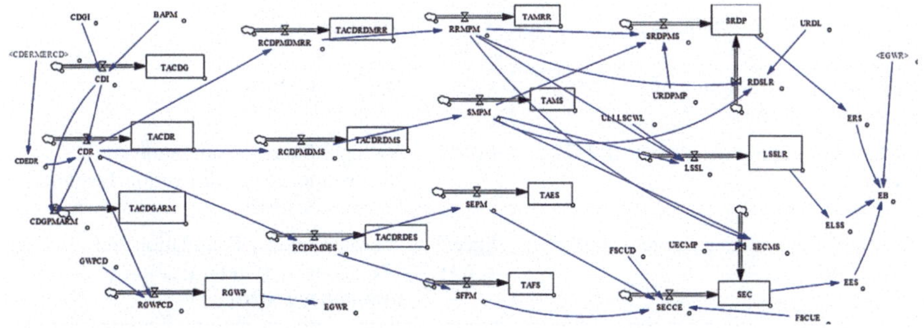

Fig. 7.4 The stock-flow diagram of environmental benefits subsystem

7.4 Model Simulation and Results Analysis

7.4.1 Model Parameter

An appropriate value of each model variable should be assigned before model simulation. To ensure the validity of parameters and equations in the proposed SD model, a construction project is selected in Qingdao, Shandong province, China and collected project data by a period of a month interview and survey. The project is a new frame-shear wall structured residential building. The gross floor area is about 109,894 m^2. Construction was conducted from February 2018 to March 2020 spanning 25 months.

Some model variables are constant parameters whose values remain the same during the simulation period. Their values can be obtained from the available research data and literature. Table 7.1 shows the units, values and sources of the quantitative variables in the model. There are also some values of variables need be obtained by on-site interview and project data. The parameter values of reduction measures for construction stage. Table 7.2 shows the parameter values in the model.

Table 7.1 Quantitative variable values

Variable	Value	Unit	Source
Building area of the project monthly	4395.7	m^2	Interview
CO_2 generation index	0.304	t/m^2	Interview
Unit resource depletion of landfilling	7.78 × 10–19	1/t	Ding et al. (2018)
Unit resource depletion potential of materials production	2.001 × 10–17	1/t	Ding et al. (2018)
Unit land loss from landfills space by construction waste landfill	0.68	m^3/t	Ding et al. (2018)
Global warming potential of CO_2	1	–	Mao et al. (2013)
Unit energy consumption of materials production	0.5593	kgce/t	Ding et al. (2018)
Factor of standard coal of unit diesel	1.2677	kgce/L	National Energy Foundation and Management Standardization Technical Committee (2020)
Factor of standard coal of unit electricity	0.1229	kgce/KWH	National Energy Foundation and Management Standardization Technical Committee (2020)

Table 7.2 Model parameter values for reduction measures

Reduction measures	Initial value
CO_2 emission reduction management investment	0.05
CO_2 emission reduction training	0.16
Environmental awareness	0.388
CO_2 emission reduction culture within construction institute	0.13
Worker's attitude to CO_2 emission reduction	0.18
Emission reduction management on-site	0.31
Local sourcing materials	0.08
Operation efficiency of construction equipment improve	0.267
Materials recycle and reuse on-site	0.35
Low carbon materials	0.15
Layout planning on-site	0.12

Many formulas in the system dynamics model involve nonlinear functions, which are specified in an analytical manner. Therefore, nonlinear functions (table function) of system dynamics can be used to represent dynamic correlation between variables.

7.4.2 Model Validation

7.4.2.1 Boundary-Adequacy Test

Boundary-adequacy test is performed for examining whether the variable and feedback contained in the model are sufficient to describe the problem and meet the predetermined purpose (Wang 2009). The model boundary was defined according to literature (Mao et al. 2013). It is found that each of variables has a significant impact in evaluating the CO_2 emissions reduction management benefits by examining all the variables in the SD model.

7.4.2.2 Dimension Consistency Test

Dimension consistency test examines whether the variable dimensions in the model equations are consistent (Wang 2009). Vensim has a function to automatically verify dimensions after defining the measurement units of all variables. The model has been verified for dimension consistency using Vensim.

7.4.2.3 Structure Verification Test

The purpose of structure verification test is to evaluating the model logic and its consistency with actual situation (Qudrat-Ullah and Seong 2010). The model structure was developed according to the causal-loop diagram in Figs. 7.2 and 7.4 which were based on a comprehensive literature review and CO_2 emissions reduction management practice. Therefore, the model structure is logical and closely consistent with the actual situation.

7.4.2.4 Extreme Condition Test

This test examines whether each equation in the model is meaningful under the extreme conditions (Ding et al. 2016). An example exemplifying the extreme condition test is employed to show the relationship between building area of project monthly (BAPM) and the total amount of CO_2 generated after reduction management (TAGARM). The impact of BAPM on TAGARM is chronologically examined by changing the value of BAPM from 0 m^2/month to 5000 m^2/month and observing how BAPM influences the model behaviors as shown in Fig. 7.5. The result show that

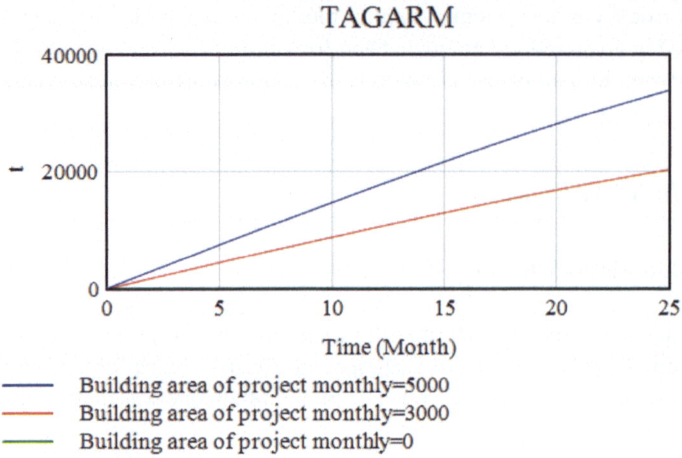

Fig. 7.5 Extreme condition test results

when the building area of project monthly equals to 0 m^2/month, the total amount of CO_2 generated after reduction management is 0 ton. When the building area of project monthly is increased, so is the total amount of CO_2 generated after reduction management. The result indicates that the model variables respond consistently. The other variables in the model were similarly and successfully tested.

7.4.2.5 Sensitivity Test

The model sensitivity test is used to analyze the values of some key variables and monitor model behavior (Ding et al. 2018). For example, by changing the value of local sourcing materials, the effect of CO_2 emissions reduction can be observed. The range of local sourcing materials is between 0 and 1 and the variation covered [0, 0.25, 0.5, 0.75, 1] as shown in Fig. 7.6. The figure emphasizes that the implementation of local sourcing materials can improve the effect of CO_2 emissions reduction.

7.4.2.6 Model Behaviors Verification Test

The model behaviors verification test is used to compared the actual data with simulation data (Pan et al. 2021). The project data of the total amount of CO_2 generated since February 2018 to March 2020 collected by interviews were used for the model behaviors verification test. The total amount of CO_2 generated was input into a Reference Mode. After the model simulation was completed, actual data and simulation data are compared as shown Fig. 7.7. The result indicates that the model as expected.

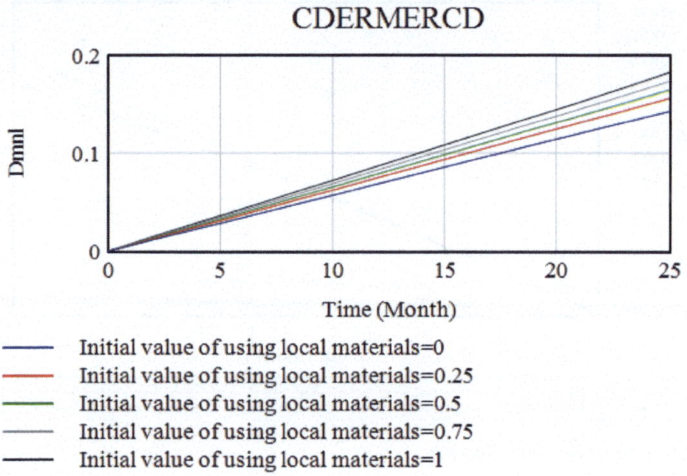

Fig. 7.6 Sensitivity test

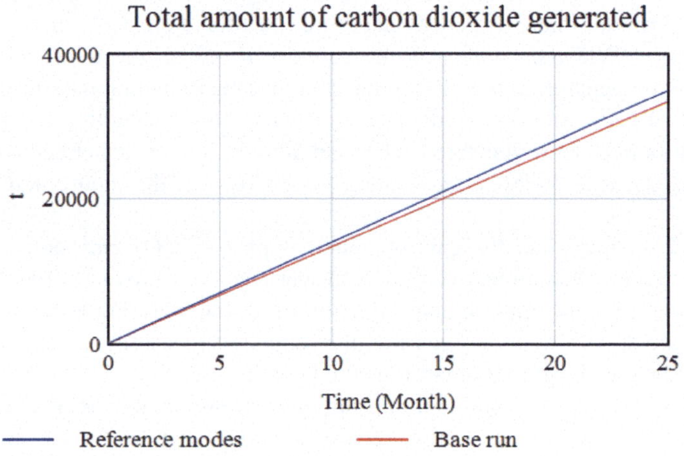

Fig. 7.7 Model behaviors verification test

7.5 Simulation Results and Discussions

7.5.1 Simulation Results

Figure 7.8 shows that the total amount of CO_2 generated, total amount of CO_2 reduction and total amount of CO_2 generated after reduction management. It can be seen that the total amount of CO_2 generated, total amount of CO_2 reduction and total amount of CO_2 generated after reduction management are increasing during

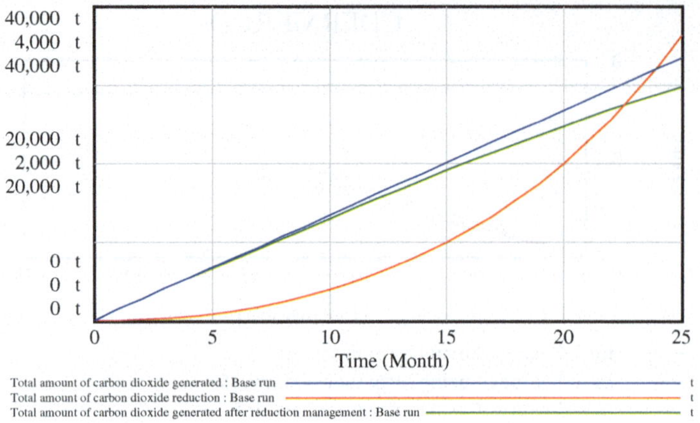

Fig. 7.8 CO_2 generation and reduction

simulation periods. This study indicates that the implementation of CO_2 emissions reduction measures at the construction stage can significantly reduce CO_2 generated.

Table 7.3 reveals that CO_2 generated before and after taking CO_2 emissions reduction measures. When no reduction measures were taken, the total of amount of carbon dioxide generation is 33407.3 ton. After taking reduction measures, the total of amount of CO_2 emissions reduction is 3620.51 ton, accounting for 10.84% of the total amount of CO_2 generated. The total amount of CO_2 generated after reduction management is 29786.8, accounting for 89.16% of the total amount of CO_2 generated.

Figure 7.9 shows that the environmental impact of CO_2 emissions reduction. Saving of resource depletion potential, land space saving from landfilling reduction, saving of energy consumption and reduction of global warming potential shows a sharp upward trend. It can be seen that the reduction measures at the construction stage can achieve significant environmental benefits.

Table 7.4 describes that simulation result of environmental benefits of CO_2 emissions reduction. It is provided comprehensive and intuitive index for decision-making in assessing the CO_2 emissions reduction benefits at the construction stage.

Table 7.3 Comparison of CO_2 emissions before and after taking reduction measures

Variable	The total amount of CO_2 generated	The total amount of CO_2 emissions reduction	The total amount of CO_2 generated after emissions reduction management
Production (ton)	33,407.3	3620.51	29,786.8
Percentage (%)	100	10.84	89.16

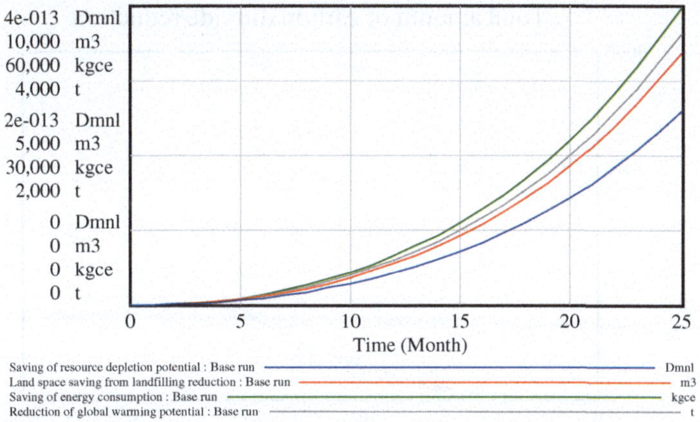

Fig. 7.9 Environmental benefits

Table 7.4 Environmental benefits of CO_2 emissions reduction

Benefit index	Improved quantity after reduction
	Value units
Saving of resource depletion potential	2.59 × 10–13 1
Land space saving from landfilling reduction	8434.28 m^3
Saving of energy consumption	59,718.5 kgce
Reduction of global warming potential	3620.51 t

7.5.2 Scenario Analysis

In the scenario analysis, the key variables, "low-carbon materials", materials recycle and reuse" and operation efficiency of construction equipment improve" are selected.

Scenario 1 (S1): low-carbon materials. The scenario analyzes impact on the total amount of CO_2 emissions reduction and environmental benefits with the increasing amounts of low-carbon materials.

In the case, the initials value of low-carbon materials variable is 0.15 (Base run), 0.5 (S2) and 0.7 (S3) respectively. By increasing the amounts of low-carbon materials, the total amount of CO_2 emissions reduction and environmental benefits can be monitored. Figure 7.10 shows the simulation results. When the initials value of low-carbon materials was adjusted to 0.5, the total amount of CO_2 emissions reduction was 4880.84 ton, accounting for 14.61% of the total amount of CO_2 generated. when the initial value of low-carbon materials was adjusted to 0.7, the total amount of CO_2 emissions reduction was 5514.84 ton, accounting for 16.50% of the total amount of CO_2 generated. When the initial value of low-carbon materials variable is changed, the improvement of environmental benefits and the total amount of CO_2 emissions reduction are consistent.

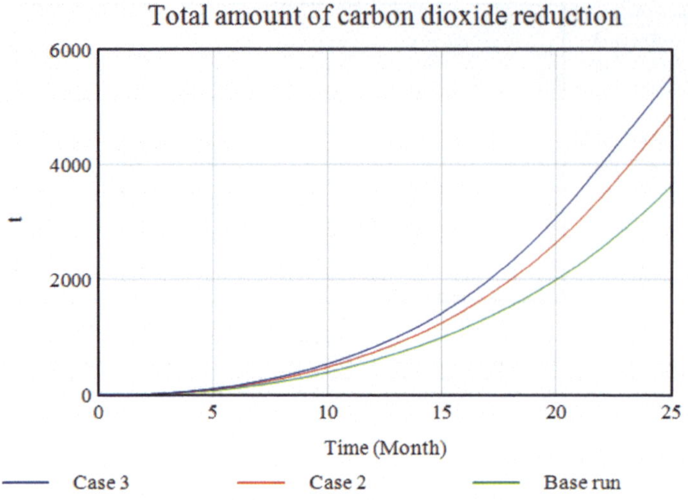

Fig. 7.10 Simulation result of the total amount of CO_2 emissions reduction under S1

Scenario 2 (S2): materials recycle and reuse. The scenario analyzes impact on the total amount of CO_2 emissions reduction and environmental benefits with the total amount of materials recycle and reuse.

In this study, the initial value of materials recycle and reuse was 0.35 (Base run), 0.5 (S2), 0.7 (S3) respectively. Figure 7.11 shows the simulation results. When the initials value of materials recycle and reuse was adjusted to 0.5, the total amount of CO_2 emissions reduction was 4160.42 ton, accounting for 12.45% of the total amount of CO_2 generated. When the initial value of materials recycle and reuse was adjusted to 0.7, the total amount of CO_2 emissions reduction was 4908.64 ton, accounting for 14.69% of the total amount of CO_2 generated. When the initial value of materials recycle and reuse variable is changed, the improvement of environmental benefits and the total amount of CO_2 emissions reduction are consistent.

Scenario 3 (S3): operation efficiency of construction equipment improved. The scenario analyzes impact on the total amount of CO_2 emissions reduction and environmental benefits with the operation efficiency of construction equipment improve.

In this study, the initial value of operation efficiency of construction equipment improve was 0.35 (Base run), 0.5 (S2), 0.7(S3) respectively. Figure 7.12 shows the simulation results. When the initials value of operation efficiency of construction equipment improve was adjusted to 0.5, the total amount of CO_2 emissions reduction was 4180.51 ton, accounting for 12.51% of the total amount of CO_2 generated. When the initial value of operation efficiency of construction equipment improve was adjusted to 0.7, the total amount of CO_2 emissions reduction was 4671.33 ton, accounting for 13.98% of the total amount of CO_2generated. When the initial value of operation efficiency of construction equipment improve variable is changed,

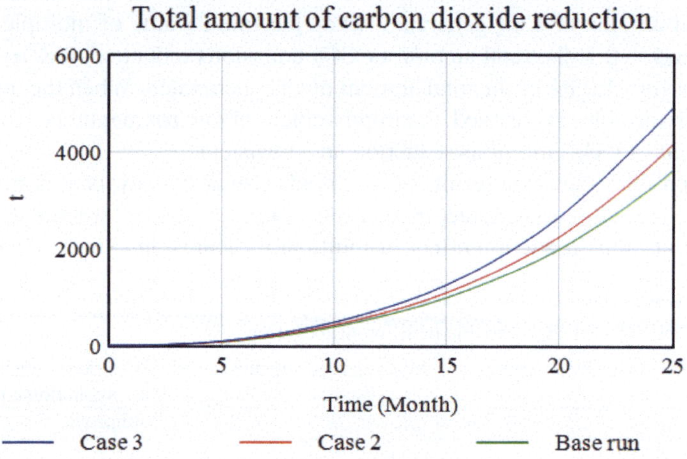

Fig. 7.11 Simulation result of the total amount of CO_2 emissions reduction under S2

the improvement of environmental benefits and the total amount of CO_2 emissions reduction are consistent.

Scenario 4 (S4): low-carbon materials, materials recycle and reuse and operation efficiency of construction equipment improve were simultaneously changed. This scenario analyzes impact on the total amount of CO_2 emissions reduction and environmental benefits with multiple variables. Table 7.5 shows the value of each variable. The simulation results of scenario analysis are shown in Fig. 7.13 and Table 7.6. When the initials value of multiple variables was adjusted to 0.5, the total amount of CO_2 emissions reduction was 5815.55 ton, accounting for 17.41%

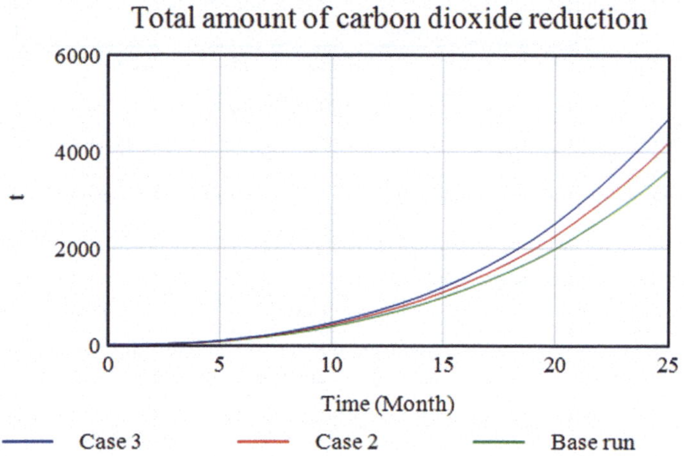

Fig. 7.12 Simulation result of the total amount of CO_2 emissions reduction under S3

of the total amount of CO_2 generated. when the initial value of multiple variables was adjusted to 0.7, the total amount of CO_2 emissions reduction was 7031.76 ton, accounting for 21.05% of the total amount of CO_2 generated. When the initial value of multiple variables is changed, the improvement of environmental benefits and the total amount of CO_2 emissions reduction are consistent.

Based on the simulation results of four kinds scenario analysis, it is summarized that single CO_2 emissions reduction measure is limited and differential. Comparing the CO_2 emissions reduction effect of multiple measures and single measure, it is

Table 7.5 The value of each variable in different scenarios

Scenario	Low-carbon materials	Materials recycle and reuse	Operation efficiency of construction equipment improve
Base run	0.15	0.35	0.267
Case 2	0.5	0.5	0.5
Case 3	0.7	0.7	0.7

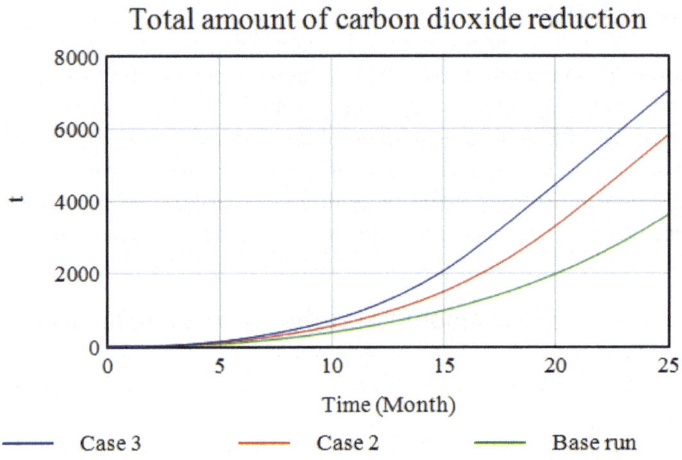

Fig. 7.13 Simulation result of the total amount of CO_2 emissions reduction under S4

Table 7.6 Multivariable scenario simulation result

Scenario 4	Base run			Case2		Case3	
	Total amount of CO_2	Reduced amount	Improved amount	Reduced amount	Improved amount	Reduced amount	Improved amount
Total amount of CO_2	33407.3	3620.51	10.84%	5815.55	17.41%	7031.76	21.05%

found that the reduction effect of multiple measures is more significant than that of any single measure. Therefore, if the best environmental benefits of CO_2 emissions reduction are achieved, multiple reduction measures must be implemented simultaneously. By the model, every project participant will acknowledge the benefit of CO_2 emissions reduction management not merely easing the alarming climate change but also bring good environmental benefits.

7.6 Conclusion

There is a general lack of approaches that enables us to holistically consider the dynamic and interdependences of variables within CO_2 emissions reduction system of construction project for complex CO_2 emissions reduction management. Therefore, this paper developed a system dynamic model for environmental benefits assessment of CO_2 emissions reduction management at the construction stage and took a residential building as an example was simulated. In order to ensure that the model is behaviorally and structurally reliable and valid, this paper carried out six kinds typical tests. The purpose of the causal-loop diagram and stock-flow diagram is to identifying key variables and describing their interrelationships.

Simulation results shows that the total amount of CO_2 generated was 33,407.3 ton, and the total amount of CO_2 emissions reduction was 3620.51 ton, accounting for 10.84% of the total amount of CO_2 generated at the construction stage. Therefore, the carbon emissions reduction management at the construction stage can bring about the following environmental benefits: the total resource consumption saving 2.59E − 13, land space saving 8434.28m^3, the total energy saving 59,718.5 kgce and reduction of global warming potential 3620.51 kg$CO_{2\text{-}e}$. scenario analysis shows that there are priorities to achieve better environmental benefits of the CO_2 emissions reduction management. In three kinds single measure, it is found that the CO_2 emissions reduction effect of low-carbon materials is best, followed by materials recycle and reuse, and the worst reduction effect is operation efficiency of construction equipment improve. As the environmental benefits of multiple measures outweigh those of single measure, primary attention should be paid to combination of different measures and achieve better and rapid environmental benefits. With the accelerated process of urbanization, the consumption of materials and energy has become increasingly increase. Therefore, effective CO_2 emissions reduction management can greatly reduce resource consumption and alleviate the increasingly severe global warming problem.

The proposed model structure in this paper was based on literature review and on-site data. The focus was to reflect the general mechanism of CO_2 emissions reduction management and their environmental benefits. This model can dynamically simulate how different measures influence environment and it can not only help construction professionals determine the optimal CO_2 emissions reduction strategy, but also provide reference for project managers to evaluate the CO_2 emissions reduction results and environmental benefits of the project. Therefore, other studies can use the model structure to simulate the environmental benefits of CO_2 emissions reduction management but the model parameter values should be adjusted to suite the concrete project data. This research focus on the environmental benefits of CO_2 emissions reduction management at the construction stage but cannot fully reveal the effect of CO_2 emissions reduction. As the architectural design have great impacts on CO_2 emissions reduction and single stage oriented analysis for either design or construction stage can possibly break the interactive factors across. Therefore, the core of future studies can simultaneously research the dynamic interaction of factors at the design and construction stage and simulate environmental benefits of two stage.

Acknowledgements The author(s) disclosed receipt of the following financial sup-port for the research, authorship, and/or publication of this article: this work was supported by the Shandong Provincial Key Research and Development Program of China (Soft Science program) (2021RKY06105); Education and Teaching Research Program of Shandong Province (General Program) (2021JXY076); Shandong Provincial Science and Technology Innovation in Emergency Management Program of China.

Appendix 1: The Description of CO_2 Emissions Reduction Management Subsystem

No.		Variable name
1	CDERMERCD	Carbon dioxide emissions reduction management efforts to reduce carbon dioxide
2	VCDERMERCD	Variation of carbon dioxide emissions reduction management efforts to reduce carbon dioxide
3	EB	Environmental benefits
4	EEB	Effect of environmental benefits
5	CDERMI	Carbon dioxide emissions reduction management investment
6	IVCDERMI	Initial value of carbon dioxide emissions reduction management investment
7	ECDERMI	Effect of carbon dioxide emissions reduction management investment
8	LCM	Low-carbon materials
9	IVLCM	Initial value of low-carbon materials
10	ELCM	Effect of low-carbon materials
11	LSM	Local sourcing materials
12	IVLSM	Initial value of local sourcing materials
13	ELSM	Effect of local sourcing materials
14	CDERT	Carbon dioxide emissions reduction training
15	IVCDERT	Initial value of carbon dioxide emissions reduction training
16	ECDERT	Effect of Carbon dioxide emissions reduction training

(continued)

(continued)

No.		Variable name
17	EA	Environmental awareness
18	IVEA	Initial value of environmental awareness
19	EEA	Effect of environmental awareness
20	CDERCWCI	Carbon dioxide emissions reduction culture within construction institute
21	IVCDERCWCI	Initial value of carbon dioxide emissions reduction culture within construction institute
22	ECDERCWCI	Effect of carbon dioxide emissions reduction culture within construction institute
23	WACDER	Worker's attitude to carbon dioxide emissions reduction
24	IVWACDER	Initial value of worker's attitude to carbon dioxide emissions reduction
25	EWACDER	Effect of worker's attitude to carbon dioxide emission reduction
26	ERMOS	Emissions reduction management on-site
27	IVERMOS	Initial value of emissions reduction management on-site
28	EERMOS	Effect of emissions reduction management on-site
29	MRRO	Materials recycle and reuse on-site
30	IVMRRO	Initial value of materials recycle and reuse on-site
31	EMRRO	Effect of materials recycle and reuse on-site
32	OECEI	Operation efficiency of construction equipment improve
33	IVIOECE	Initial value of improvement of operation efficiency of construction equipment
34	EIOECE	Effect of improvement of operation efficiency of construction equipment
35	LPO	Layout planning on-site
36	IVLPO	Initial value of layout planning on-site
37	ELPO	Effect of layout planning on-site

Appendix 2: The Description of Environmental Benefits Subsystem

No.	Abbreviation	Variable name
1	CDDR	Carbon dioxide decrease rate
2	CDGI	Carbon dioxide generation index
3	BAPM	Building area of the project monthly
4	CDI	Carbon dioxide increase

(continued)

(continued)

No.	Abbreviation	Variable name
5	TACDG	Total amount of carbon dioxide generated
6	CDR	Carbon dioxide reduction
7	TACDR	Total amount of carbon dioxide reduction
8	CDGPMARM	Carbon dioxide generated per month after carbon dioxide reduction management
9	TACDGARM	Total amount of carbon dioxide generated after carbon dioxide reduction management
10	RCDPMDMRR	Reduce carbon dioxide per month due to materials recycle and reuse
11	TACDRDMRR	Total amount of carbon dioxide reduced due to materials recycle and reuse
12	RCDPMDMS	Reduce carbon dioxide per month due to materials substituted
13	TACDRDMS	Total amount of carbon dioxide reduced due to materials substituted
14	RCDPMDES	Reduce carbon dioxide per month due to energy saved
15	TACDRDES	Total amount of carbon dioxide reduced due to energy saved
16	RRMPM	Recycle and reuse materials per month
17	TAMRR	Total amount of materials recycled and reused
18	SMPM	Substitute materials per month
19	TAMS	Total amount of materials substituted
20	SEPM	Save electricity per month
21	TAES	Total amount of electricity saved
22	SFPM	Save fuel per month
23	TAFS	Total amount of fuel saved
24	SRDP	Saving of resource depletion potential
25	SRDPMS	Saving of resource depletion potential from material saved
26	RDSLR	Resource depletion saving from landfilling reduction
27	URDPMP	Unit resource depletion potential of material production
28	URDL	Unit resource depletion of landfilling
29	LSSLR	Land space saving from landfilling reduction
30	LSSL	Land saving from landfilling reduction per month
31	ULLLSCWL	Unit land loss from landfills space by construction waste landfill
32	SEC	Saving of energy consumption
33	SECMS	Saving of energy consumption from material saved
34	UECMP	Unit energy consumption of material production
35	SECCE	Saving of energy consumption from on-site construction equipment
36	FSCUD	Factor of standard coal of unit diesel
37	FSCUE	Factor of standard coal of unit electricity
38	RGWP	Reduction of global warming potential
39	RGWPCD	Reduction of global warming potential from carbon dioxide

(continued)

(continued)

No.	Abbreviation	Variable name
40	GWPCD	Global warming potential of carbon dioxide
41	ERS	Effect of resource saving
42	EES	Effect of energy saving
43	ELSS	Effect of land space saving
44	EGWPR	Effect of global warming potential reduction
45	EB	Environmental benefits

References

Ahn C, Lee S, Peña-Mora F, Abourizk S (2010) Toward environmentally sustainable construction processes: the US and Canada's perspective on energy consumption and GHG/CAP emissions. Sustainability 2(1):354–370

Bank LC, McCarthy M, Thompson BP, Menassa CC (2010) Integrating BIM with system dynamics as a decision-making framework for sustainable building design and operation. In: The first international conference on sustainable urbanization (ICSU 2010)

Basbagill J, Flager F, Lepech M, Fischer M (2013) Application of life-cycle assessment to early stage building design for reduced embodied environmental impacts. Build Environ 60:81–92

Cao X, Li X, Zhu Y, Zhang Z (2015) A comparative study of environmental performance between prefabricated and traditional residential buildings in China. J Clean Prod 109:131–143

Chau CK, Hui WK, Ng WY, Powell G (2012) Assessment of CO_2 emissions reduction in high-rise concrete office buildings using different material use options. Resour Conserv Recycl 61:22–34

Cheng C, Pouffary S, Svenningsen N, Callaway M (2008) The Kyoto protocol, the clean development mechanism, and the building and construction sector. UNEP SBCI, Paris

Clevenger CM, Khan R (2014) Impact of BIM-enabled design-to-fabrication on building delivery. Pract Period Struct Des Constr 19(1):122–128

Ding Z, Yi G, Tam VW, Huang T (2016) A system dynamics-based environmental performance simulation of construction waste reduction management in China. Waste Manage 51:130–141

Ding Z, Zhu M, Tam VW, Yi G, Tran CN (2018) A system dynamics-based environmental benefit assessment model of construction waste reduction management at the design and construction stages. J Clean Prod 176:676–692

González MJ, Navarro JG (2006) Assessment of the decrease of CO_2 emissions in the construction field through the selection of materials: practical case study of three houses of low environmental impact. Build Environ 41(7):902–909

Guggemos AA, Horvath A (2006) Decision-support tool for assessing the environmental effects of constructing commercial buildings. J Archit Eng 12(4):187–195

Guo R, Zhao Y, Shi Y et al (2017) Low carbon development and local sustainability from a carbon balance perspective. Resour Conserv Recycl 122:270–279

Inyim P, Rivera J, Zhu Y (2015) Integration of building information modeling and economic and environmental impact analysis to support sustainable building design. J Manag Eng 31(1):A4014002

Lai JHK, Yik FWH, Man CS (2012) Carbon audit: a literature review and an empirical study on a hotel. Facilities

Lewis P, Leming M, Rasdorf W (2012) Impact of engine idling on fuel use and CO_2 emissions of nonroad diesel construction equipment. J Manag Eng 28(1):31–38

Lim DB, Son S, Kim S, Park J (2019) Life cycle cost and CO_2 emission simulation models of apartment building projects using system dynamics. Indoor Built Environ 28(3):310–321
Ma P (2017) Analysis and research on environmental benefits of green buildings from the perspective of carbon emission, master dissertation. Tianjin University
Mao C, Shen Q, Shen L, Tang L (2013) Comparative study of greenhouse gas emissions between off-site prefabrication and conventional construction methods: two case studies of residential projects. Energy Buildings 66:165–176
Naqi A, Jang JG (2019) Recent progress in green cement technology utilizing low-carbon emission fuels and raw materials: a review. Sustainability 11(2):537
National Energy Foundation and Management Standardization Technical Committee (2020) General rules for calculation of comprehensive energy consumption. National Energy Foundation and Management Standardization Technical Committee, Beijing
Ng WY, Chau CK (2015) New life of the building materials-recycle, reuse and recovery. Energy Procedia 75:2884–2891
Ortiz O, Pasqualino JC, Díez G, Castells F (2010) The environmental impact of the construction phase: an application to composite walls from a life cycle perspective. Resour Conserv Recycl 54(11):832–840
Ozkaymak M, Ceylan MA, Okutan H et al (2017) CO_2 emission during the combustion of Orhaneli lignite coal. World J Eng
Padilla-Rivera A, Amor B, Blanchet P (2018) Evaluating the link between low carbon reductions strategies and its performance in the context of climate change: a carbon footprint of a wood-frame residential building in Quebec, Canada. Sustainability 10(8):2715
Pan X, Li M, Xu H, Guo S, Guo R, Lee CT (2021) Simulation on the effectiveness of carbon emission trading policy: a system dynamics approach. J Oper Res Soc 72(7):1447–1460
Papachristos G, Jain N, Burman E, Zimmermann N, Mumovic D, Davies M, Edkins A (2020) Low carbon building performance in the construction industry: a multi-method approach of project management operations and building energy use applied in a UK public office building. Energy and Buildings 206:109609
Qudrat-Ullah H, Seong BS (2010) How to do structural validity of a system dynamics type simulation model: the case of an energy policy model. Energy Policy 38(5):2216–2224
Rasmussen FN, Birkved M, Birgisdóttir H (2020) Low-carbon design strategies for new residential buildings–lessons from architectural practice. Archit Eng Design Manage 16(5):374–390
Sadok RH, Belaribi NB, Mazouzi R et al (2022) Life cycle assessment of cementitious materials based on calcined sediments from Chorfa II dam for low carbon binders as sustainable building materials. Sci Total Environ 826:154077
Saghafi MD, Teshnizi ZSH (2011) Recycling value of building materials in building assessment systems. Energy Buildings 43(11):3181–3188
Sandanayake M, Zhang G, Setunge S, Luo W, Li CQ (2017) Estimation and comparison of environmental emissions and impacts at foundation and structure construction stages of a building—a case study. J Clean Prod 151:319–329
Sbci UNEP (2009) Buildings and climate change: summary for decision-makers. United Nations Environmental Programme, Sustainable Buildings and Climate Initiative, Paris, 1–62
Szamocki N, Kim MK, Ahn CR et al (2019) Reducing greenhouse gas emission of construction equipment at construction sites: field study approach. J Constr Eng Manag 145(9):05019012
Teng Y, Li K, Pan W, Ng T (2018) Reducing building life cycle carbon emissions through prefabrication: evidence from and gaps in empirical studies. Build Environ 132:125–136
Thompson BP, Bank LC (2010) Use of system dynamics as a decision-making tool in building design and operation. Build Environ 45(4):1006–1015
UN Environment Programme (2016) Emissions gap report
Wang Q (2009) System dynamics. Shanghai University of Finance and Economics Press, Shanghai, pp 240–246
Yuan H, Wang J (2014) A system dynamics model for determining the waste disposal charging fee in construction. Eur J Oper Res 237(3):988–996

Zhang X, Zheng R (2020) Reducing building embodied emissions in the design phase: a comparative study on structural alternatives. J Clean Prod 243:118656

Zhang X, Platten A, Shen L (2011) Green property development practice in China: costs and barriers. Build Environ 46(11):2153–2160

Zhou Z, Zhang S, Wang C, Zuo J, He Q, Rameezdeen R (2016) Achieving energy efficient buildings via retrofitting of existing buildings: a case study. J Clean Prod 112:3605–3615

Chapter 8
Covid-19 Impacts on ASEAN Aquatic Environment: Towards Mitigation Actions for Impending Resilience on Aquaculture and Fisheries in Brunei Darussalam

Wanidawati Tamat, Wafi Diah, Cristalina Jalil Marsal, Ahmed Jalal Khan Chowdhury, Anisah Syakirah Anwari, Mohd Hafiz Jamaludin, and Adibah Johari

Abstract A novel coronavirus disease called COVID-19 is a persistent problem on a global scale. Worldwide aquatic food production is affected by COVID-19, much like other sectors that produce food. Aquaculture methods and output are being negatively impacted by the COVID-19 epidemic and the ensuing lockdowns and limitations. Aquaculture is the practice of raising aquatic organisms for human consumption in aquatic environments. Aquaculture operations cannot be carried out without the involvement of humans. Because humans have manipulated freshwater, brackish water, and marine ecosystems to develop aquaculture, different aquaculture methods are the results of interactions between humans and the environment. The

---

W. Tamat (✉) · W. Diah
Department of Fisheries, Ministry of Primary Resources and Tourism, Brunei Muara 1728, BT, Brunei Darussalam
e-mail: wanidawati.tamat@fishereis.gov.bn

W. Diah
e-mail: wafi.diah@fisheries.gov.bn

C. J. Marsal · A. J. Khan Chowdhury · A. S. Anwari · M. H. Jamaludin · A. Johari
Faculty of Agriculture, Universiti Islam Sultan Sharif Ali (UNISSA), KM 33, Jalan Tutong, Kampung Sinaut, Tutong TB1741, Brunei Darussalam
e-mail: cristalina.jalil@unissa.edu.bn

A. J. Khan Chowdhury
e-mail: ahmed.chowdhury@unissa.edu.bn

M. H. Jamaludin
Faculty of Agrobased Industry, University Malaysia Kelantan, 17600 Jeli, Kelantan, Malaysia

A. Johari
Geology Group, Faculty of Science, Universiti Brunei Darussalam, Jalan Tungku Link, Gadong, Bandar Seri Begawan BE1410, Brunei Darussalam

D. Jones and A. J. Khan Chowdhury (eds.), *Environmental Resilience in the Pandemic Years 2020–2021*, Environmental Science and Engineering,
https://doi.org/10.1007/978-3-031-47757-7_8

COVID-19 pandemic in Southeast Asia has had a significant negative impact on the fisheries and aquaculture industry, particularly on the socioeconomic circumstances of the stakeholders, such as fishers, fish farmers, traders, and consumers. In fact, the COVID-19 pandemic had adverse effects on the global aquatic environment affecting both artisanal and commercial fishing as well as aquaculture (species, markets, and farm financial capacity), producers supplying the food services industry (hotels, restaurants, and catering), working conditions along the supply chain (particularly for women and other vulnerable workers), and fisheries management practices. However, this review paper has discussed comprehensively the influence of COVID-19 on the aquatic environments towards various post-pandemic mitigating methods for the aquaculture and fishing industries in the ASEAN region, with a distinct emphasis on Brunei Darussalam.

Keywords COVID-19 · Aquaculture and fisheries management · Supply chain · Aquatic environment · Mitigation

8.1 Introduction

8.1.1 Covid-19 Pandemic Environmental Consequences

COVID-19 pandemic is a severe threat for mankind having an extensive socio-economic impact. A strange pneumonia-like illness was reported from the Chinese province of Wuhan in the latter part of 2019; it is thought to be related to the wet market in the province (Xu et al. 2020). The World Health Organization's emergency committee was informed right away of the abrupt disease epidemic (WHO). On January 30, 2020, the WHO declared COVID-19 a "Public Health Emergency of International Concern (PHEIC)" (WHO 2020a) due to several threats to humans, including COVID-19's high infectivity, asymptotic carrier, mild clinical symptoms, lack of human immunity from previous infections, and uncertain incubation time (Bouey 2020; Rithchie et al. 2021). By keeping all these in mind WHO has also suggested various pandemic risk reduction strategies like social distancing, partial and complete lockdowns, mass quarantines, extensive travel bans, and disruption of transportation systems, which could have a direct impact on local and global socio-political relations and economic growth (Dente and Hashimoto 2020). Implementation of worldwide lockdown offers an unprecedented and unique opportunity to study the impact of anthropogenic activities on the natural ecosystem and wild flora and fauna (Bates et al. 2020). Apart from the human suffering and economic loss caused by the pandemic, it has offered various environmental gains including remarkable improvement of the water quality index. As many countries across the globe underwent lockdown to prevent the spread of COVID-19, many wild animals were seen to enjoy a pollution-free, peaceful environment without any human interference (Corlett et al. 2020).

8.2 Impacts of Covid-19 on the Aquatic Environment

The aquatic environment consists of a continuum of fresh water aquatic ecosystems from the upstream to the downstream estuary and coastal area towards mariculture an deep sea fishing area. Besides, massive marine environment, numerous lakes, and wetlands in the flood plains and/or river basins also contributed immensely to the lotic-lentic water body complex. With the rapid economic development, especially in developing countries, many of these water bodies are subjected to environmental stressors associated with anthropogenic activities and climate change, resulting in water quality deterioration, harmful algal blooms, loss of productivity, and loss of biodiversity (Yusoff et al. 2021).

Several current concerns to freshwater ecosystems, such as the introduction of exotic species, pollution, fragmentation, flow restriction, habitat loss, and exploitation, are being modulated by COVID-19 lockdown (Cooke et al. 2021; Corlett et al. 2020). Due to COVID-19 shutdown, invasive species that have proliferated across different locations have undergone changes. The introduction of alien species linked to ballast water exchange, air transportation, the movement of fresh goods, and recreational activities is typically reduced because of reduced travel (Early et al. 2016; Edward et al. 2020).

For instance, decreasing shipping traffic among all ports worldwide results in a large decline in species invasion related to the interchange of ballast water (Sardain et al. 2019). Assessing the current state of water pollution, it has been noted that industrial activity closures, both full and partial, may have resulted in a decrease in the flow of pollutants such as heavy metals, nutrients, and harmful compounds to water bodies (Sharma et al. 2021).

Around 10% of the ocean's surface is occupied by the coastal ecosystem, which is crucial to maintaining the ocean's biogeochemical cycle in general and the carbon cycle. Coastal areas typically receive a high load of anthropogenic materials, which can occasionally cause eutrophication and changes to aquatic ecology (Seitzinger et al. 2020). The deep-sea ecology is significantly impacted by the introduction of anthropogenic contaminants to the coastal environment. According to several studies, the main cause of coastal/marine water pollution is human-induced pollutant input. The COVID-19 pandemic lockdown, however, fully stopped or severely curtailed tourist, fishing, industrial activity, agriculture, and aquaculture practices. From this point forward, as expected, the degree of coastal pollution (Lotliker et al. 2021).

Nevertheless, the COVID-19 pandemic, which was brought on by the SARS-CoV-2 coronavirus, changed the ecological makeup of aquatic habitats and had an impact on aquatic food sources worldwide population's socioeconomic system and supply network. The aquatic environment's water quality, wild fishing stocks, and biodiversity were said to have improved because of decreased human activity during the epidemic (Degnarain 2020). In fact, there is no evidence at this time that COVID-19 can be spread between humans through food. According to the World Health Organization (WHO 2020a) and the International Organization for Animal Health, the COVID-19 pandemic is sustained by human-to-human transmission rather than by international trade in animals and animal products. Fish and fisheries products could become infected if handled by people who don't practice adequate hygiene and have COVID-19 (FAO 2020a).

The COVID-19 pandemic may be a scourge for humanity, but it may also be a boon for certain aquatic habitats that are already stressed by intensive exploitation and growing pollution. This worldwide halt may present one of the unexpected possibilities for the resuscitation of fish species, 90% of which are over or fully exploited (FAO 2020b). The second world war's suspension of commercial fishing was historically seen to cause a marked increase in fish landings across the world (Beare et al. 2010). As a result, under the current situation, an unintentional COVID-19 lockout could speed up the recovery of fish stocks and the habitats where they live, such as coral reefs, mangroves, and river bottoms. Although COVID-19 does not affect aquatic life, the pandemic has impacted or at the very least slowed down the bulk of aquaculture practices worldwide (FAO 2020c). Many farmers were compelled to temporarily stop raising fish during the epidemic or severely curtail their aquaculture practices. Since COVID-19 is now a serious human health concern, human–environment interactions in aquaculture are at risk from the COVID-19 pandemic because land-based aquaculture is so heavily impacted by pandemic anxieties (WHO 2020b; Nelson and Surachetpong 2020; Jalal 2020, 2021).

Many aquaculture practices around the world have been impacted by the epidemic. Aquaculture is very susceptible to the pandemic, which has caused a few problems with the supply of inputs and is now affecting grow-out operations. The COVID-19 epidemic and the ensuing physical barriers, lockdowns, and physical distancing have all affected aquaculture practices, in the first instance, by limiting input supply (FAO 2021). The appearance of COVID-19 showed both its positive and negative impacts on the aquatic environment (Table 8.1).

8.2.1 Aquaculture and Fisheries Sectors in ASEAN Region During the COVID-19 Pandemic

Aquaculture operations have been hampered by an erratic and inadequate supply of inputs during the lockdown, which has complicated seafood marketing efforts and risked the production of aquaculture goods. Global aquaculture productivity has been damaged by COVID-19-related lockdowns and limitations. According to the most recent projections, the COVID-19 pandemic will cause a nearly 2% decrease in worldwide aquaculture productivity in 2020 (FAO 2021). Due to issues with input, production, marketing, and export, it has been alleged that COVID-19 has severely damaged China's aquaculture industry (Chun 2020). In several Asian nations, it has been observed that COVID-19 negatively affects aquaculture production (Lala-Pritchard et al. 2020; Islam et al. 2021; Alam et al. 2022). As a result, COVID-19 expanded swiftly throughout the world and had a significant negative impact on businesses, especially the fisheries and aquaculture sectors, as well as on people's

Table 8.1 Impacts of COVID-19 on the aquatic environment

Components/ elements	Impacts/results	Reasons	References
Water quality (Physico-chemical parameters)	Decrease in total solids, less	Less human activities and decreased discharges. Turbidity levels decreased by 25% due to reduction in human activities	Praveena and Aris (2021) Sun et al. (2020)
	Improvement in suspended	Decreased of SPM by 15.9%	Yunus et al. (2020)
	Increased water transparency	Reduction in water-based activities due to lockdown	Braga et al. (2020)
	Decrease in nutrients less agro-based industries, less nutrient rich	Less agro-based industries, less nutrient rich	Bandala et al. (2021)
	Decrease of some heavy metal concentrations in surface and ground waters	Decrease in industrial discharges	Selvam et al. (2020)
	Improvement of water quality index (based on DO, BOD, COD, pH and NH3-N) in rivers and lakes	Significant reduction in industrial and agriculture activities and human encroachment. Closure of industrial and tourism activities	Najah et al. (2021)
Water as a Source to Transmit Viruses	Surface waters in less-or-untreated sewage were found to include SARS-CoV-2 identified viruses	Increase of COVID-19 cases and evidence its presence in waste waters	Bandala et al. (2021) Liu et al. (2020) Albastaki et al. (2021)

Source Yusoff et al. (2021)

quality of life. In addition to generating cash and creating jobs, the fisheries and aquaculture sector is crucial for nutrition and food security. Prior to the virus outbreak, the global fishery and aquaculture industry produced 401 billion USD for human and non-human consumption in 2018.

Fish and fisheries goods were widely produced in the Southeast Asian region and traded not only on the international markets but also on domestic and local markets. Because of this, COVID-19 spread swiftly over the world and had a significant negative impact on businesses, notably the fisheries and aquaculture sectors, as well as on people's quality of life. Food security and nutrition, as well as revenue production and job creation, are all significantly influenced by the fisheries and aquaculture sector (FAO 2020d; SEAFDEC 2020a).

In 2018, the global economy received USD 401 billion from the production of fisheries and aquaculture around the world for human and non-human consumption before the viral epidemic. Fish and fisheries products were produced in large quantities in many Southeast Asian nations, and they were marketed not only on the international markets but also on the domestic and local markets. In fact, the world's overall fishery, the region contributed about 46.5 million t (21.9%) in total (FAO 2020d).

There is currently no proof that COVID-19 can be transmitted to people through food (Bondad-Reantaso et al. 2020). The COVID-19 pandemic is perpetuated by human-to-human transmission, not by international traffic in animals and animal products, according to the World Health Organization (WHO) and the International Organization for Animal Health (OAH). When handled by those who have COVID-19 and don't exercise good hygiene, fish and fisheries goods may get contaminated (FAO 2022). However, the COVID-19 pandemic had a negative impact on the world's fisheries and aquaculture industry, which included fishing (on both an artisanal and commercial scale), aquaculture (species, markets, and farm financial capacity), producers of food services (hotels, restaurants, and catering), and working conditions along the supply chain (particularly for women and other underrepresented groups).

The COVID-19 pandemic has had significant negative effects on the fisheries and aquaculture sector, particularly on the socioeconomic circumstances of the stakeholders, such as fishers, fish farmers, traders, and consumers, as the suspension of food services along the supply chain places great strain on them not only with regard to the production aspect but also with regard to marketing and trade in the ASEAN region (Kaewnuratchadasorn et al. 2020; Jalal 2020, 2021).

Additionally, because of their limited capacity to lessen the negative effects of crises, small-scale fishing households in the region were already vulnerable. Noteworthy, the aquaculture production of Brunei Darussalam, Indonesia, Malaysia, Myanmar, Singapore, and Thailand during 2018–2020 is shown in Table 8.2.

For Brunei Darussalam, although the production decreased from around 1248.25t (USD 10.33 million) in 2018 to around 973.99 t (USD 7.98 million) in 2019, there was a significant increase with around 3501.38 t (USD 24.26 million) in 2020 which was due to the significant increase in inland aquaculture production (FAO 2020e).

On the contrary, the capture fishery production of Brunei Darussalam, Indonesia, Malaysia, Myanmar, Singapore, and Thailand during 2018–2020 as stated in Table 8.3. For Brunei Darussalam, the total capture (only marine) fishery production slightly increased from 13,566t (USD 55.34 million) to 13,725t (USD 55.55 million) but decreased to 13,074t (USD 55.19 million).

Table 8.2 Aquaculture production of Brunei Darussalam, Indonesia, Lao PDR, Malaysia, Myanmar, Singapore, and Thailand by quantity (t) and value (USD million) from 2018 to 2020

Country		Quantity (million t)			Value (USD)		
		2018	2019	2020	2018	2019	2020
Brunei Darussalam	Brackish water	417.64	371.57	339.24	2.06	1.76	2.51
	Inland	724.30	590.90	3,045.78	4.89	4.13	19.56
	Marine	106.31	11.52	116.36	0.81	0.08	1.13
Indonesia	Brackish water	3,537,737.38	3,630,880.53	–	–	–	–
	Inland	9,268,216.17	879,033.21	–	–	–	–
	Marine	2,965,851.74	2,962,394.10	–	–	–	–
Lao PDR	Inland		67,000	122,000		33.58	33.97
Malaysia	Brackish water	290,195.00	307,181.00	302,807.25	565.28	615.64	582.70
	Inland	101,270.00	104,602.00	97,210.32	171.35	190.37	190.19
Myanmar	Total	1,130,350[a]	1,121,350[b]	–	–	–	–
Singapore	Brackish water	395.00	369.00	–	6.31	5.86	–
	Inland	854.00	1,014.00	–	6.11	7.86	–
	Marine	4,453.00	4,448.00	–	20.99	22.43	–
Thailand	Brackish water	495,022.00	536,466.00	514,600.00	2,242.68	2,361.01	2,235.09
	Inland	425,840.00	427,330.00	405,656.00	832.04	829.94	786.61

8.2.2 Aquaculture and Fisheries Production in Brunei Darussalam

Brunei Darussalam is in the northwestern part of Borneo. The country has a total land area of 5765 km^2 and a coastline of 269 km long fronting the South China Sea. The continental shelf area is about 8600 km^2 and the exclusive economic zone has an area of 5614 km^2. The fisheries subsectors in Brunei Darussalam include capture fisheries, aquaculture, and seafood processing. Capture fisheries are composed of small-scale and commercial fisheries. For the purpose of fisheries management, the fishing area is divided into four zones, namely: Zone 1 (0–3 nm), Zone 2 (3–20 nm), Zone 3 (20–45 nm), and Zone 4 (45–200 nm). The Department of Fisheries under the Ministry of Primary Resources and Tourism is the fisheries authority of Brunei Darussalam. Aquaculture is the human cultivation of aquatic organisms in waters and aquaculture operation is not possible without direct human involvement. In Brunei Darussalam, different aquaculture practices are the outcomes of human–environment interactions, as humans have created aquaculture through the manipulation of freshwater, brackish water, and marine habitats (SEAFDEC 2021; DOF 2021).

Table 8.3 Capture fishery production of Brunei Darussalam, Indonesia, Lao PDR, Malaysia, Myanmar, Singapore, and Thailand by quantity (t) and value (USD million) from 2018 to 2020

Country		Quantity (million t)			Value (USD)		
		2018	2019	2020	2018	2019	2020
Brunei Darussalam	Marine	13,566	13,725	13,074	55.34	55.55	55.19
Indonesia	Marine	6,696,336	7,813,550	1,549,963[a]	–	–	–
	Inland	659,517.89	736,972.00	148,990.00	–	–	–
Lao PDR	Inland		67,000	122,000	–	15.97	16.16
Malaysia	Marine	1,452,862	1,455,446	1,383,299	2,726	2,765	2,506
	Inland	6,089.00	5,569.00	5,625.14	29.97	90.17	83.18
Myanmar	Marine	3,249,700[b]	3,264,830[c]	–	–	–	–
	Inland	1,600,050	1,608,060	–	–	–	–
Singapore	Marine	1,310	1,418	–	9.08	9.00	
Thailand	Marine	1,392,931	1,410,665	1,492,878	1,938	2,133	1,933
	Inland	143,825	116,465	123,256	276	226	233

Note *Fiscal year 8.2018–8.2019; **Fiscal year 8.2019–8.2020
Source SEAFDEC (2020a, 2020b)

Capture Fisheries, Aquaculture, and Seafood Processing are the three (3) subsectors that make up Brunei Darussalam's Fisheries Industry. A comparison of each subsector's GDP contribution to the fisheries sector is shown in Fig. 8.1. This indicates that aquaculture has grown by 14% over the past five years, which is the biggest growth. Corresponding to that, the processing sector saw a 10% gain over the previous five years. On the other hand, the proportion of marine capture, which constitutes the largest share in fisheries, has decreased from 2015 to 2020 by 25%.

It is evident from the annual production figures for the aquaculture industry that the increase in value and volume of production was not progressive. After increasing from 2015 to 2017, production fell in 2018 and 2019, but it sharply surged in 2020, rising more than twice as much as the year before, from 10.64 million BND to 32.35 million BND, as shown in Figs. 8.2 and 8.3.

By analyzing the same output numbers from the standpoint of the commodity composition produced by aquaculture, as seen in Fig. 8.4, marine shrimp has continuously been the sector's largest contributor, followed by marine fish. The graph's steep spike in 2020 is the most striking feature, which has already been mentioned in earlier parts. According to the breakdown, marine shrimp output increased fivefold in a year, which is what caused the sector's production value to significantly increase.

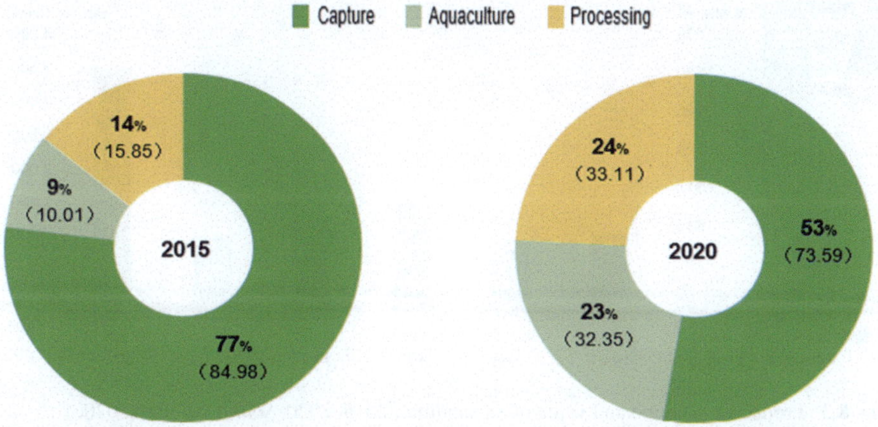

Fig. 8.1 Brunei Fisheries subsector production share, 2015 and 2020. *Source* Department of Economic Planning and Statistics, Ministry of Finance and Economy MOFE (2020a)

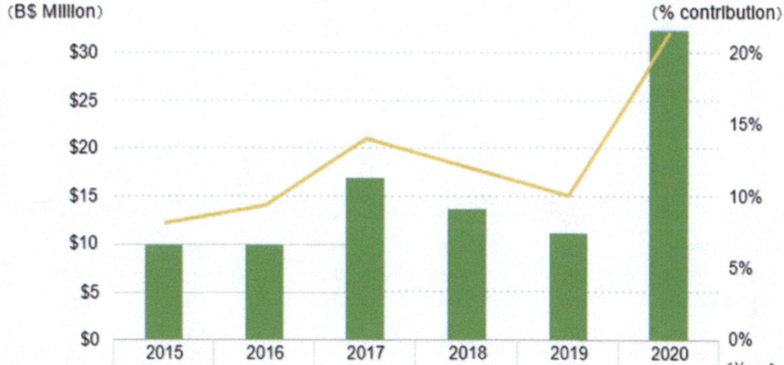

	2015	2016	2017	2018	2019	2020
Production from aquaculture (BND Million)	10.01	9.89	16.7	13.77	10.64	32.35
% Contribution from aquaculture to fisheries industry's Production	9.03%	10.38%	14.92%	12.57%	10.62%	23.27%

Fig. 8.2 Production value of aquaculture and contribution to fisheries, 2015–2020. *Source* Department of Economic Planning and Statistics, Ministry of Finance and Economy MOFE (2020b)

8.2.3 Production of Fish and Seafood

In recent years, Brunei's entire fishery production has increased, rising from $110.84 million BND in 2015 to $139.05 million BND in 2020 (Table 8.4). The output of aquaculture, one of the three subsectors of the fisheries industry—capture, aquaculture, and processing contributed significantly to the increase from $10.01 million BND (9% in share) in 2015 to $32.35 million BND (23% in 2020). (See Fig. II.6).

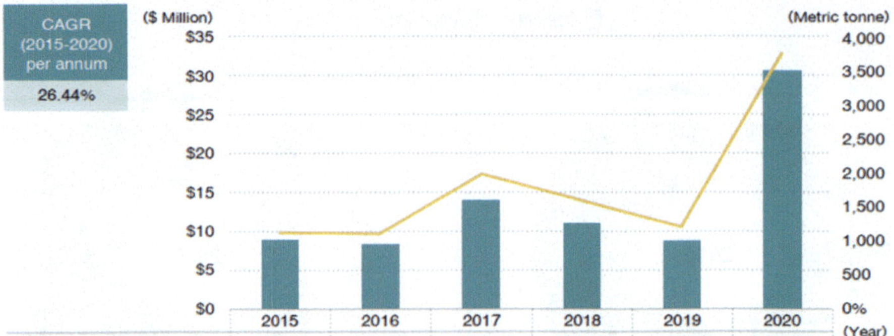

	2015	2016	2017	2018	2019	2020
Quantity (metric tonne)	983.57	948.86	1632.18	1248.25	973.99	3,501.38
Value (BND million)	10.01	9.89	16.7	13.77	10.64	32.35

Fig. 8.3 Production volume and value of aquaculture, 2015–2020. *Source* MOFE (2020c)

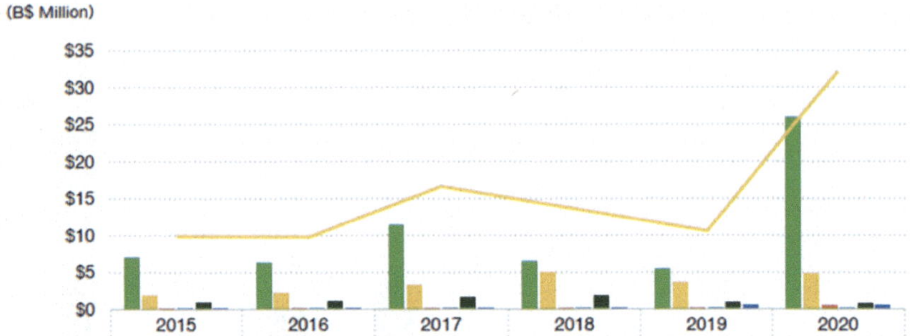

	2015	2016	2017	2018	2019	2020
Marine Shrimp	7.09	6.36	11.56	6.52	5.51	26.08
Marine Fish	1.85	2.26	3.31	5.1	3.67	4.82
Freshwater Fish	0.07	0.06	0.05	0.09	0.09	0.27
Crab	0	0	0	0.01	0	0.02
Marine Shrimp Fry	0.92	1.03	1.7	1.76	0.84	0.64
Fish Fry	0.07	0.18	0.08	0.3	0.54	0.53
Total	10.01	9.89	17.7	13.77	10.64	32.35

Fig. 8.4 Aquaculture production by type of species, 2015–2020. *Source* Department of Economic Planning and Statistics, Ministry of Finance and Economy. MOFE (2020c)

The growth in fisheries production appears to be driven by the growth in aquaculture production, which has climbed from 9% in 2015 to 23% in 2020.

As it is evidently showed in Fig. 8.5, the aquaculture production in the last decade has been particularly substantial in terms of growth rate. Although the number declined in 2018 and 2019, the production appears to have recovered in 2020 with the highest production quantity at all times.

Table 8.4 Total fisheries industry production between 2015 and 2020

Year	2015	2016	2017	2018	2019	2020
Value (BND million)	$110.84	$95.34	$111.90	$109.52	$100.15	$139.05

Source DOF (2021)

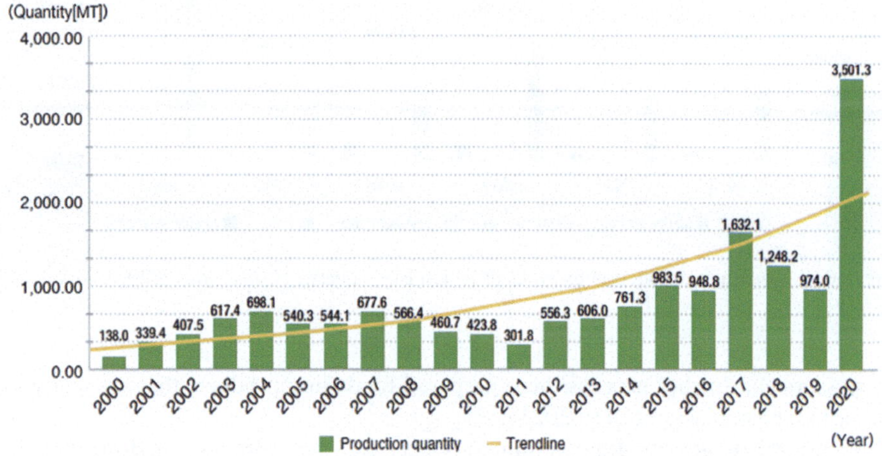

Fig. 8.5 Production of aquaculture industry between 2000 and 2020. *Source* DOF (2021)

Regarding the specific species that are raised in Brunei, Fig. 8.6 shows that blue shrimp output was significantly higher in 2018 when compared to other species. Even while Brunei produces other high-value fish as well, like grouper and barramundi, their volume is still insignificant in comparison to that of blue shrimp. Given its bulk, grouper appears to be contributing much in terms of values.

The high-value fishes can effectively contribute to the production values till 2020 even if the volume is relatively modest, as is seen with grouper fish. Diseases in aquaculture continue to be a major biosecurity concern that must be addressed to support wholesome and healthy aquaculture.

8.2.4 Aquatic Products Processing

Apart from marine capture and aquaculture, a new business called seafood processing has emerged. After being gathered by marine capture and aquaculture, fish and other seafood items can either be sold directly to the local market, undergo packaging and preparation for export, or be transferred to facilities for processing. Whole fish or shellfish are processed into a variety of products, including fish fillets, fish steaks, frozen goods, parts of breaded fish, canned goods, and smoked goods. Secondary

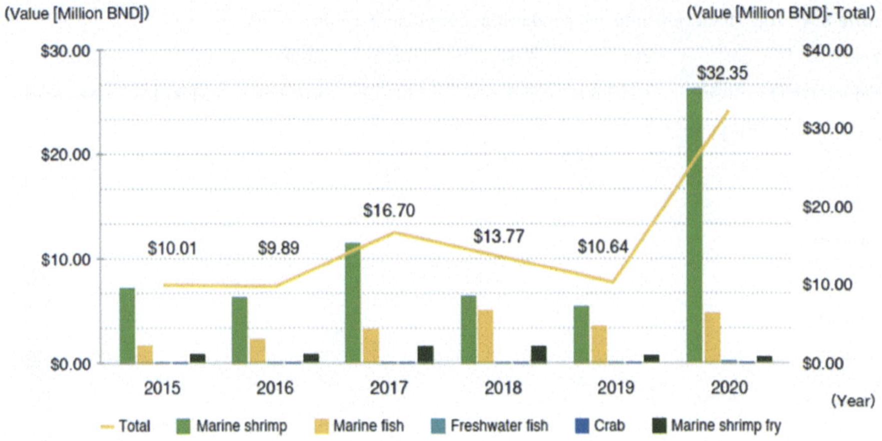

Fig. 8.6 Aquaculture production by types of fish in Brunei between 2015 and 2020. *Source* DOF (2021)

processors can further process these prepared foods into products that are ready to eat or to heat and serve (DOF 2021).

Figure 8.7 depicts the total volume produced by each subsector of Brunei's fisheries having a significant volume difference from marine capture, the processing business comes in second. In recent years, the processing industry's output volume has varied. After falling between 2014 and 2016, it rose during the next three years, reaching a value of 4115.9 metric tonnes in 2019.

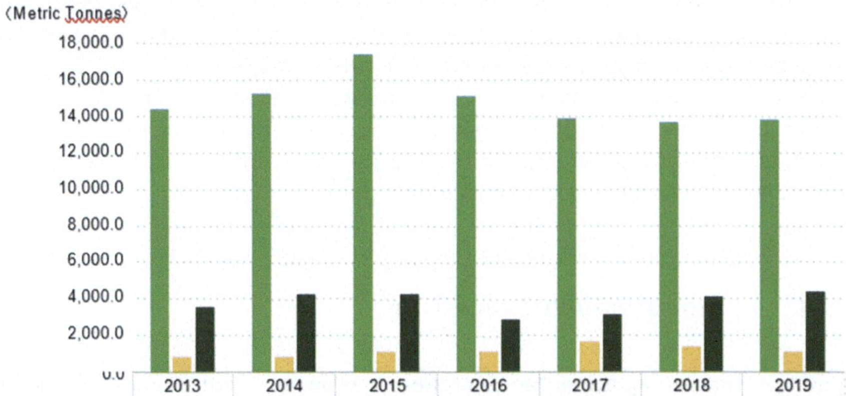

	2013	2014	2015	2016	2017	2018	2019
Capture Industry	14,320.3	15,100.4	17,333.3	14,966.1	13,795.6	13,565.6	13,724.5
Aquaculture Industry	606.0	761.3	983.5	948.8	1,632.1	1,248.2	973.9
Processing Industry	3,325.7	4,068.7	3,983.7	2,645.9	2,963.4	3,852.6	4,115.9

Fig. 8.7 Production of fisheries per subsector in metric tonnes, 2013–2019. *Source* Ron and Haruko (2021), DOF (2021)

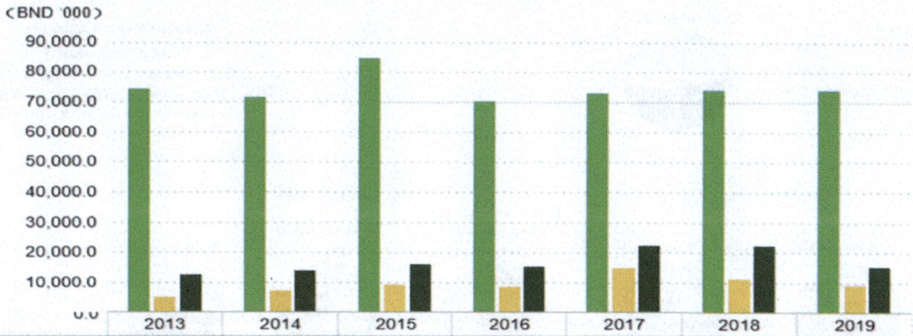

	2013	2014	2015	2016	2017	2018	2019
Capture Industry	74,608.2	71,500.8	84,982.3	70,113.9	73,037.3	73,805.8	74,072.1
Aquaculture Industry	4,872.2	6,992.6	9,055.6	8,688.1	14,919.9	11,713.0	9,268.4
Processing Industry	12,168.0	14,202.0	15,826.0	15,329.0	22,190.0	21,940.0	15,440.1

Fig. 8.8 Value of fisheries per subsector in metric tonnes, 2013–2019. *Source* Department of Economic Planning and Statistics, Ministry of Finance and Economy MOFE (2020c)

On the contrary, Fig. 8.8 shows the production of Brunei fisheries per subsector in terms of value in BND in the past seven years. While the processing industry is still far second to marine capture, there is a noticeable difference in the trend where processing production has increased steadily compared to the decline in capture fisheries. The value of the aquaculture industry increased from 2013 to 2015 before slightly declining in 2016. It then continued to increase in 2017 where it peaked at 22,190 million BND. But then, this has declined in the following two years with the value dropping to 15,440.1 million BND in 2019 (DOF 2021; Ron and Haruko 2021).

8.2.5 Overview Domestic Value Chain in Brunei Darussalam

To manufacture and transport the product to the market and consumers, aquaculture production necessitates a variety of actions. As a result, the value chains are crucial in establishing an environment where all parties participating in the process relevant Departments, Industry, and stakeholders can benefit from one another. By analyzing each stage of the production process, it aims to increase the value of the finished product through various effective divisions including management, market information systems, and transportation.

In Fig. 8.9: The Department of Fisheries under the Ministry of Primary Resources and Tourism (MPRT) and Brunei Economic Development Board have recognized Brunei's native fisheries ecosystem. It demonstrates how Brunei's fishing sectors are intertwined and how fish and seafood products are produced along a value chain from pre-production to distribution. The growth of fry, fingerlings, feeding, storage, and certification are all included in this process. While Fig. 8.8, captures the basic flows of the Industry and Some Spin-off Industries that are created in the Production Process.

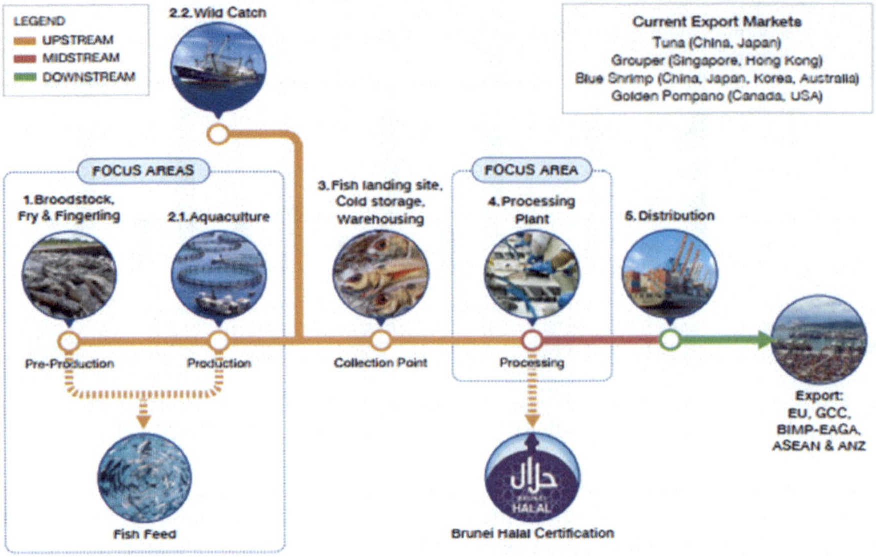

Fig. 8.9 Generalized image of the fisheries ecosystem. Specific value chains for specific areas has been developed for strategic plan for DOF, Brunei Darussalam. *Source* Ron and Hakuko (2021), DOF (2021)

Figure 8.9 shows a generalized image of the fisheries ecosystem, specific value chains for specific areas have been developed from the strategic plan of the Department of Fisheries (DOF). While Figs. 8.10 and 8.11 depict the current value chain and the process of the aquaculture and seafood processing industries in Brunei. It captures the basic flows of the industry and some spin-off industries that are created in the process of production.

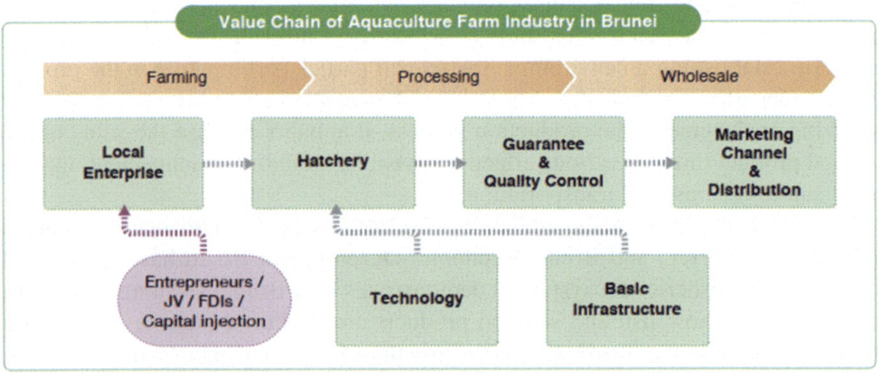

Fig. 8.10 Depict the current value chain and the process of the aquaculture and seafood processing industry in Brunei. *Source* Ron and Haruko (2021), DOF (2021)

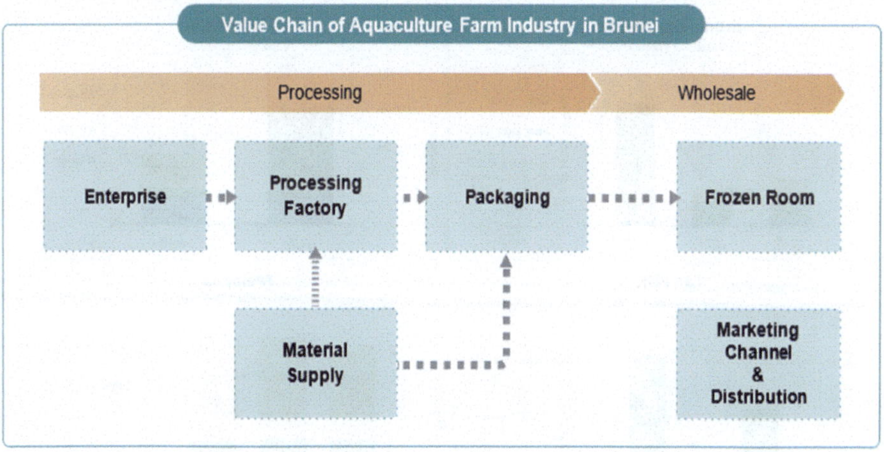

Fig. 8.11 Value chain and spin-off industries of seafood processing in Brunei. *Source* Ron and Haruko (2021), DOF, BD (2021)

The number of fishers and fish farmers in 2021 was 2195 and 446, respectively. In 2019, the gross domestic product from fisheries was 0.32% while the fish available for consumption was 41.20 kg/capita. For total production, marine capture fisheries contribute around 94% and aquaculture contributes around 6%. For Brunei Darussalam, although the aquaculture production decreased from around 1248.25 t (USD 10.33 million) in 2018 to around 973.99 t (USD 7.98 million) in 2019, there was a significant increase with around 3501.38 t (USD 24.26 million) in 2020 which was due to the significant increase in inland aquaculture production. Figure 8.12 has shown the aquaculture production in ASEAN region during 2018–2020.

On the contrary the capture fisheries sector shows in Brunei Darussalam: the total capture (only marine) fishery production slightly increased from 13,566 t (USD 55.34 million) to 13,725 t (USD 55.55 million) but decreased to 13,074 t (USD 55.19 million). Figure 8.13 has shown the Fisheries production in ASEAN region during 2018–2020 (FAO 2020a).

8.2.6 Effects on Covid-19 on Aquaculture Management in Brunei Darussalam and ASEAN Region

There were no changes in Brunei Darussalam's number of operational aquaculture farms, access to fish farms, length of the aquaculture cycle from rearing to harvest, volume of production, volume of ornamental fish production, farm gate prices of market-size cultured species, income from aquaculture activities, liquidity, and income from other fish farming endeavors. The operators had to pay increased logistic expenses for import and export, despite reliable logistics and access to

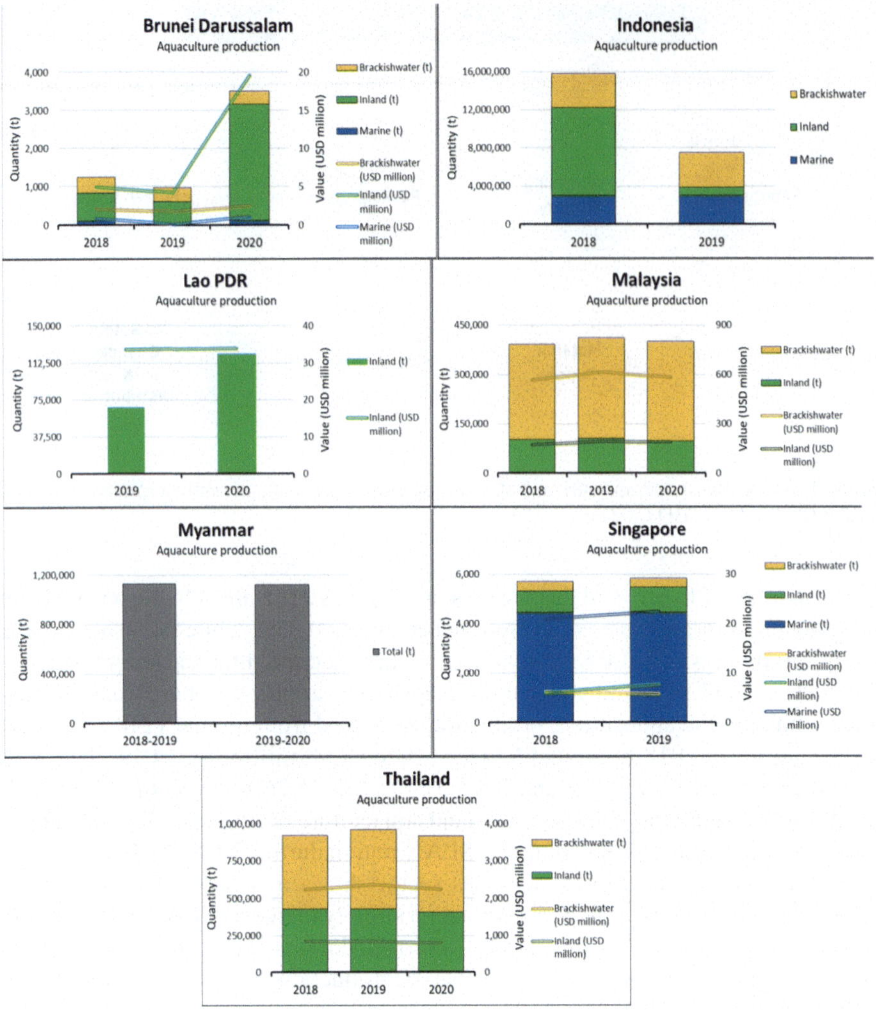

Fig. 8.12 Aquaculture production of Brunei Darussalam, Indonesia, Lao PDR, Malaysia, Myanmar, Singapore, and Thailand by quantity (t) and value (USD million) from 2018 to 2020. *Source* FAO Fishery and Aquaculture Information and Statistics Service Fisheries Country Profile: Brunei Darussalam, 2022

domestic markets. However, because of the epidemic and dependency on imports, the cost of inputs, notably raw materials, had soared. As a result, the government granted funding to import raw materials such fish fry by charter flight while the quantity of seed production had decreased substantially (Fig. 8.14).

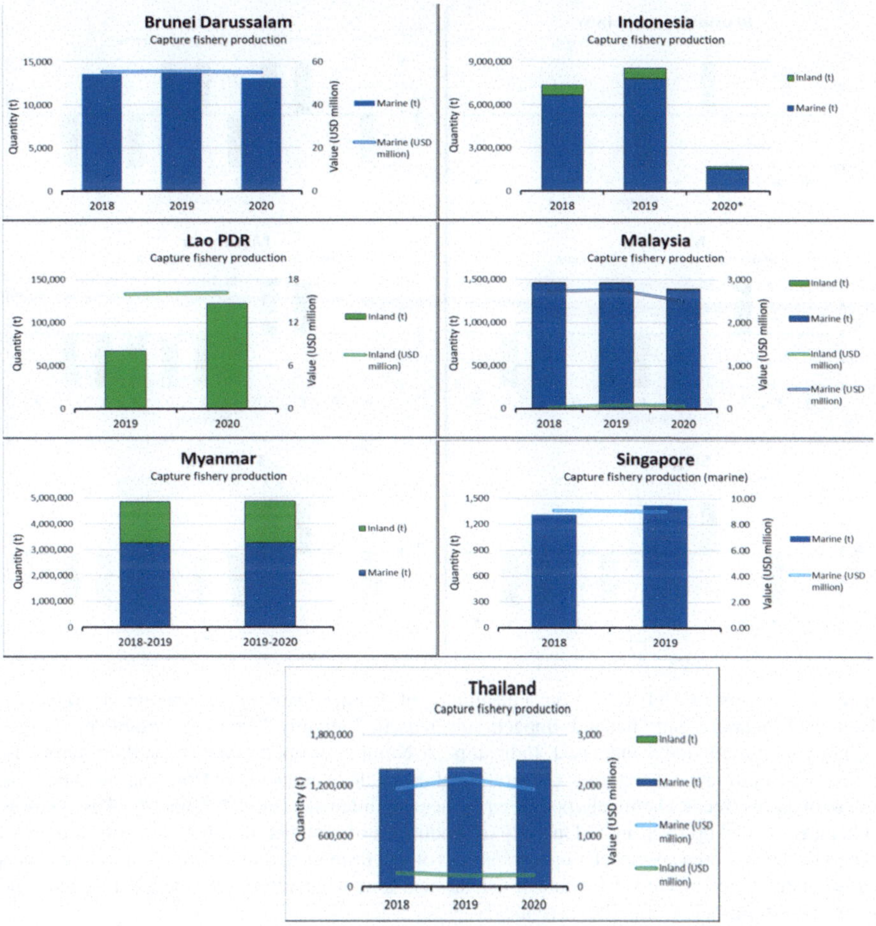

Fig. 8.13 Fisheries production of Brunei Darussalam, Indonesia, Lao PDR, Malaysia, Myanmar, Singapore, and Thailand by quantity (t) and value (USD million) from 2018 to 2020. *Source* FAO Fishery and Aquaculture Information and Statistics Service Fisheries Country Profile: Brunei Darussalam, 2022

8.3 Effects of Covid-19 on the Fisheries Management

Despite the implementation of new technologies, IUU fishing has not altered in Brunei Darussalam due to the continuous use of existing technology. However, the port State measures were carried out frequently because there were no requests from foreign fishing vessels to visit the nation's fishing port throughout the pandemic. The regular data collection for transboundary species monitoring, fish stock assessment, and shared stock monitoring were unaffected as well. In a similar way, inspection activities have dropped in inspection regions to stop the infection of fishery officials.

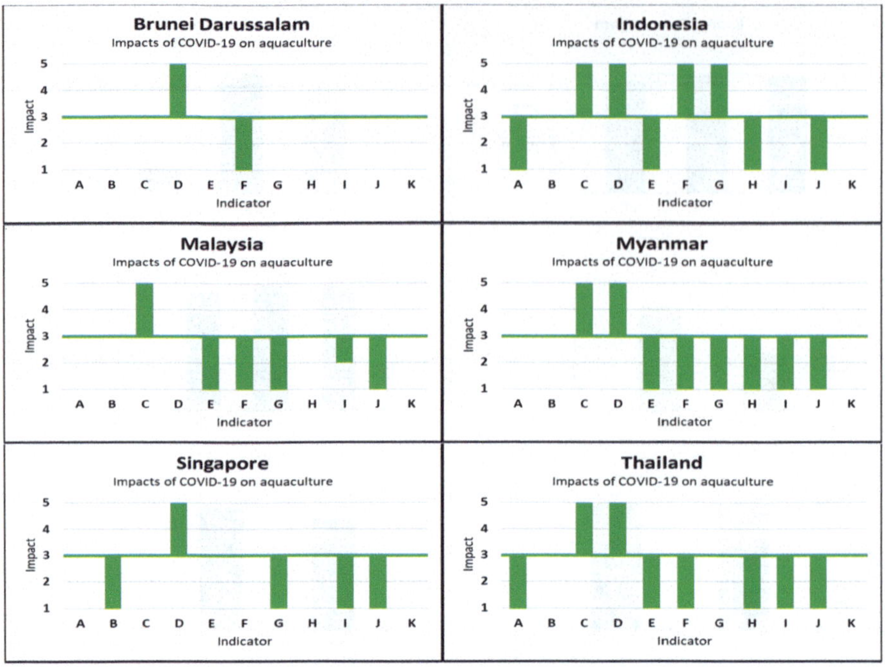

Fig. 8.14 Impacts of COVID-19 on aquaculture of Brunei Darussalam, Indonesia, Malaysia, Myanmar, Singapore, and Thailand (impact: 1-decreased; 2-slightly decreased; 3-stable/no change; 4-slightly increased; and 5-increased. Indicators: A-Number of operational aquaculture farms; B-Access of fish farmers to fish farms; C-Duration of the cycle of aquaculture from rearing to harvest; D-Cost of inputs (feeds, chemicals, power, equipment, maintenance, etc.); E-Quantity of production; F-Quantity of seed production; G-Quantity of production of ornamental fishes; H-Farm gate prices of market-size cultured species; I-Logistics/access of fish farmers to domestic/international markets (transportation, buyers, etc.); J-Income of fish farmers from aquaculture activities; K-Liquidity and income of fish farmers from other activities)

In addition, other adjustments have been made, such as the inspection of fishing vessels with a smaller staff or remote inspection via teleconference, as well as the use of other mobile applications in place of port inspection for Fisheries Management techniques in other neighboring nations of Brunei Darussalam (SEAFDEC 2020a) (Fig. 8.15).

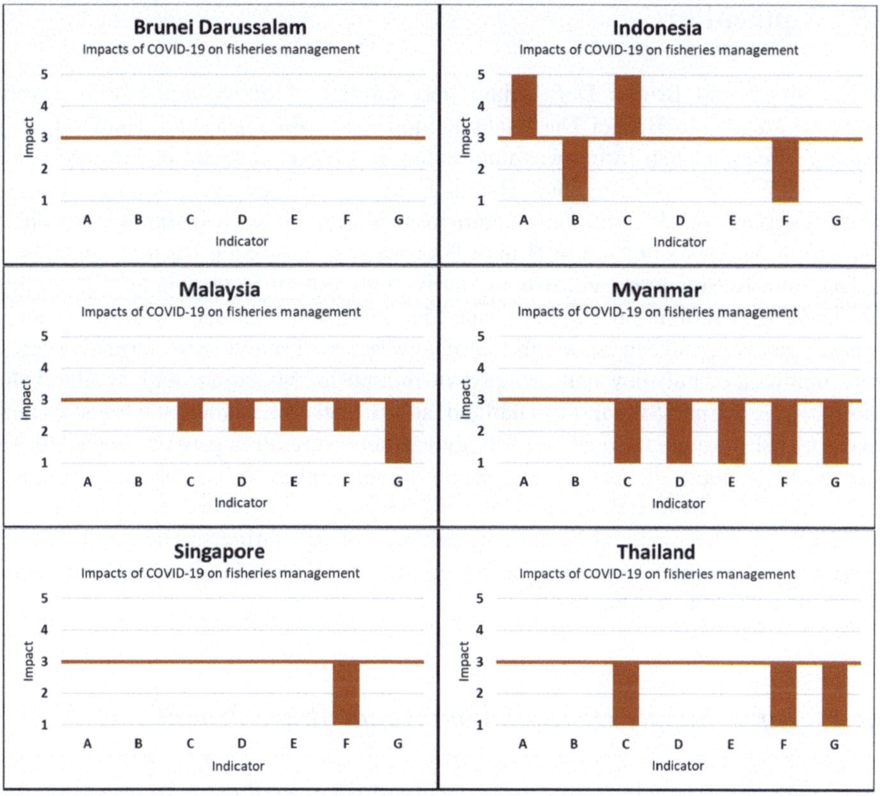

Fig. 8.15 Impacts of COVID-19 on fisheries management of Brunei Darussalam, Indonesia, Malaysia, Myanmar, Singapore, and Thailand (Impact: 1-Decreased; 2-Slightly decreased; 3-Stable/no change; 4-Slightly increased; and 5-Increased. Indicators: A-Application of innovative technologies to combat IUU fishing (GIS, remote sensing, etc.); B-Frequency of implementation of port State measures; C-Implementation of MCS; D-Conduct of regular/routine data collection for fish stock assessment; E-Monitoring of shared stocks/transboundary species; F-Conduct of physical meetings/workshops at international/regional/national levels; G-Conduct of research/project activities)

8.4 Recommendations to Mitigate the Impacts of the Post Covid-19 Pandemic on Aquaculture and Fisheries Management

A complete analysis of COVID-19's effects has been discussed to preserve the management of Brunei Darussalam's fisheries and aquaculture and its neighboring countries under ASEAN-SEAFDEC Member Countries during and after the pandemic. All interested parties must be supported by these initiatives and mitigation measures.

8.5 Aquaculture

In Singapore and Brunei Darussalam, the number of active aquaculture farms remained steady. In Brunei Darussalam, Indonesia, Malaysia, and Thailand, fish farmers' access to fish farms was unaffected, but it had reduced in Singapore. In Indonesia, Malaysia, Burma, and Thailand, the aquaculture cycle is longer than it is in Singapore and Brunei Darussalam from rearing to harvest. Input prices had gone up in Malaysia but not in Brunei Darussalam, Indonesia, Burma, Singapore, or Thailand. In Singapore and Brunei Darussalam, output levels were constant, but they had dropped in Indonesia, Malaysia, Myanmar, and Thailand. In Brunei Darussalam, Malaysia, and Singapore, the farm gate prices of market-size farmed species were unaffected, but they had dropped in Indonesia, Myanmar, and Thailand. In Lao PDR, Myanmar, Singapore, Thailand, and Brunei Darussalam, the logistics and access of fish farmers to local and foreign markets were all negatively impacted. In Brunei Darussalam, the aquaculture sector's contribution to fish farmers' incomes was consistent, but in Indonesia, Malaysia, Myanmar, Singapore, and Thailand, it had declined. The financial health and earnings of fish farmers. The liquidity and income of fish farmers from other activities were not affected in Brunei Darussalam (SEAFDEC 2020a, 2020b, 2022).

8.5.1 Input Supply Disruption on Aquaculture Production

Each stage of the supply chains for aquaculture has been directly impacted by the protection measures implemented by governments of nearly every nation in the world to prevent the spread of COVID-19. In fact, the situation is negatively affecting the supply chains for aquaculture inputs in a number of ways. In the end, the pandemic lockdowns with local, national, and international transportation halts and travel restrictions have resulted in a reduction in aquaculture input supply capacity. The effect of the COVID-19 pandemic on the supply of various aquaculture operations' inputs is a major source of concern. Brunei and other ASEAN nations' aquaculture is impacted by COVID-19 because of insufficient and/or inconsistent input supply (SEAFDEC 2020b, Nesar and Azra 2022) (Figs. 8.16 and 8.17).

COVID-19 pandemic has had a variety of effects on the fisheries and aquaculture sectors of the individual countries. Such effects may intensify either favorable or unfavorable outcomes. Therefore, the suggestions for mitigating the effects of the COVID-19 pandemic are provided below, as appropriate, for the relevant stakeholders including fishers, fish farmers, fishery workers, fishing operators, fish traders, consumers, national and local government, fisheries agencies, fisheries officers, NGOs, academe, research institutions, financial institutions, relevant non-fisheries sectors, and others. The Recommendations to Mitigate the Impacts of the COVID-19 Pandemic on Aquaculture has shown in Table 8.5.

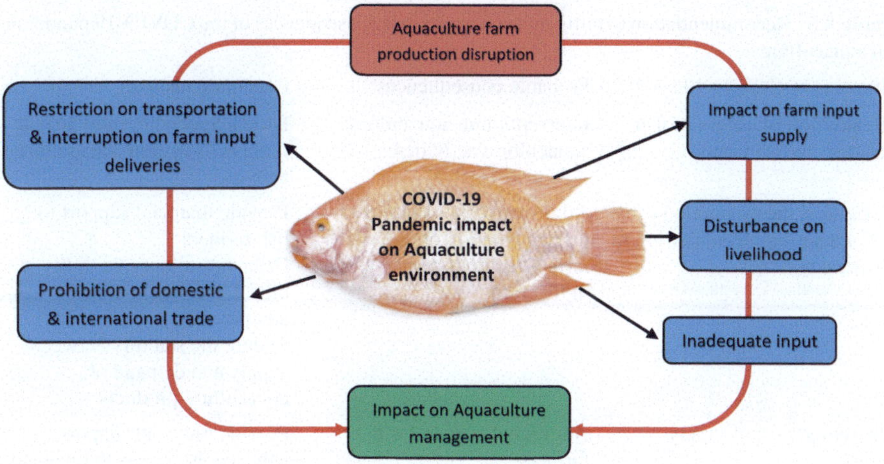

Fig. 8.16 COVID-19 lock down disruptions on aquaculture production in Brunei and other ASEAN countries. Aquaculture production and value chains in the COVID-19 pandemic. *Source* Nesar and Azra (2022), Jalal (2020)

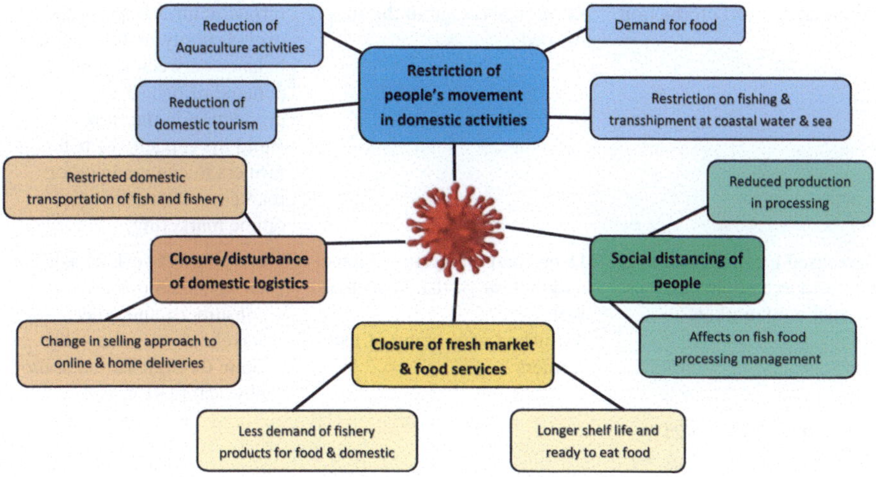

Fig. 8.17 Impacts of COVID-19 to fisheries and aquaculture. *Source* SEAFDEC (Jalal 2020, 2021)

Table 8.5 Recommendation to mitigate the impacts and consequences of the COVID-19 pandemic on aquaculture

Impacts of COVID-19	Probable consequences	Recommendations
Restriction of access of fish farmers to fish farms	Interruption in aquaculture aquaculture activities	Issue a Special Permits to fish farmers to sustain aquaculture activities
Extended the production cycle of aquaculture from rearing to harvest period	Additional cost of inputs Overpopulation of cultured farm sized commercial species in fish farm	Provide financial support to fish farmers Offer subsidies (*e.g.*, utility bills, feeds, chemicals, equipment) Endure the stability between supply and demand of aquaculture products
Increased cost of inputs	Additional burden for fish farmers	Provide financial support enhance the access to financial services (*e.g.* low-interest loans) provide subsidies (*e.g.* utility bills, feeds, chemicals, equipment)
Decreased quantity of Commercial seed production	Decrease in income of fish farmers Shortage in the supply of seeds	Develop storage infrastructures to preserve excess supply make sustain the balance between supply and demand of fish seed production in the farm Build the capacity of fish farmers to engage in other marketing strategies (e.g. online marketing)
Decreased logistics/access of fish farmers to domestic/international markets	Decreased marketing channel and income of fish farmers as well Tarnished well-being of fish farmers	• Develop network of fish farmers, fish traders, consumers, and other stakeholders in the value chain through virtual and/or physical platforms

Source SEAFDEC (2021)

8.6 Marine Capture Fisheries

8.6.1 Small-Scale Fisheries

The number of individuals involved in small-scale fishing operations had increased in Brunei Darussalam for the small-scale marine capture fisheries subsector, while it had fallen in Malaysia, Myanmar, and Thailand. There was a little drop in Myanmar and Thailand and an increase in Brunei Darussalam for the number of small-scale fishing vessels in operation. In Thailand and Brunei Darussalam, the length or frequency of fishing activity was unaffected; however, in Myanmar, it was shorter. While Malaysia and Thailand's fishing costs remained unchanged, they had increased in Brunei Darussalam and Myanmar. In Brunei Darussalam, Malaysia, and Thailand, the catch per fishing trip was unaffected; however, it had declined in Myanmar. The price of catch in wholesale markets/landing centers was stable in Brunei Darussalam and Malaysia but decreased in Thailand. The logistics/access of fishers to markets was not affected in Brunei Darussalam but had decreased in Malaysia, Myanmar, and Thailand. The income of fishers from small-scale fishing activities was not affected in Brunei Darussalam but slightly decreased in Malaysia in Myanmar and Thailand (OECD 2021).

The liquidity and income of small-scale fishers from other activities were stable in Brunei Darussalam but decreased in Myanmar. The recommendations to mitigate the impacts of the COVID-19 pandemic are summarized in Table 8.6.

8.6.2 Commercial Fisheries

For the commercial marine fisheries subsector, the duration/period of fishing activity was not affected in Brunei Darussalam but decreased in Myanmar. The cost of fishing operation had increased in Brunei Darussalam and Myanmar but there was no change in Malaysia and Thailand. [To include the issues on 1] shortage of fishing vessel crew. Mostly the crews are foreign such as Indonesia, Philippines, China. They couldn't come over to Brunei due to the travel restriction. 2]. Few operators couldn't do their vessel maintenance since couldn't travel to the shipyard maintenance which are mostly located in Miri. The amount of catch per fishing trip had not changed in Brunei Darussalam, Malaysia, and Thailand but decreased in Myanmar. The price of catch in wholesale markets/landing centers had been stable in Brunei Darussalam and Malaysia but decreased in Thailand. The transshipment at sea and cold chain systems were not affected in Myanmar and Malaysia, respectively. The access to fish ports was not affected in Malaysia but decreased in Myanmar. The capacity of cold storage facilities had been stabled in Malaysia. The liquidity and income of fishing operators from fishing had decreased in Myanmar and Thailand (Table 8.7).

8.7 Inland Fisheries

Lakes, rivers, and estuaries demonstrated improvements in their water quality and biotic life with a decrease in industrial, commercial, and other anthropogenic activity linked to COVID-19-restricted movements, demonstrating that humas are to blame for their deterioration and pollution. In comparison to the earlier time before the COVID-19 pandemic, the potential for aquatic creatures to multiply at a quicker pace was provided by the improvement in water quality and the decrease in fishing pressure (Cooke et al. 2021).

Table 8.6 Recommendations to mitigate the impacts of the COVID-19 pandemic on small-scale capture fisheries (SEAFDEC 2022)

Impacts of COVID-19	Probable consequences	Recommendations
Increasing number of persons engaged in small-scale fishing activities	Greater understanding of the value of small-scale fisheries for people who lost their jobs in cities Many small-scale fishermen are now making money from their fishing endeavors Degradation of fishery and natural resources due to increased fishing pressure	Increase locals' knowledge of the current fisheries laws and regulations among those who returned to their hometowns and engaged in small-scale fishing activities Control the number of fishing vessels operating, issue licenses or permits to fishing vessels Improve the fishing prowess and maritime safety of the locals who returned home and took up small-scale fishing activity
Increased number of small- scale fishing vessels in operation	Excess in the supply of fish and fishery products Degradation of fishery and natural resources due to increased fishing pressure	Issue for certain duration license/permit for fishing vessels to regulate the number of fishing vessels in operation Discourage indiscriminate fishing and Promote responsible fishing practices for the sustenance of marine resources
Decreased liquidity and revenue from other sources for small-scale fishermen	Deprivation of the well-being of small-scale fishers	Support financial service access (e.g. low- interest loans) Offer financial aid (e.g. subsistence allowance and supplies) Increase the capacity of small-scale fishers to engage in alternative livelihoods

Table 8.7 Recommendations to mitigate the impacts of the COVID-19 pandemic on commercial capture fisheries

Impacts of COVID-19	Possible consequences	Recommendations
Decreased duration/period of fishing activity	Decrease in the income of fishing operators and fish workers Increase in the number of unemployed fish workers Recovery of fishery and natural resources due to decreased fishing pressure	Issue a special permit to fishing operators and fish workers to sustain commercial fishing activities Facilitate the employment of local and migrant fish workers
Decreased access to fish ports	Spoilage of catch onboard fishing vessels	Upgrade the storage facilities onboard fishing vessels
Decreased liquidity and income of fishing operators from fishing	Bankruptcy of fishing operators	Build the capacity of fish workers to engage in alternative livelihoods
Decreased price of catch in wholesale markets/landing centers	Decrease in the income of fishing operators and fish workers	Establish ability of fishers to engage in other marketing strategies
Decreased amount of catch per fishing trip	Shortage in the supply of fish and fishery products	Issue a special permit to fishing operators and fish workers to sustain commercial fishing activities

Source SEAFDEC (2021)

According to Edward et al. (2020), the ASEAN riverine ecosystem's water quality parameters including turbidity, nutrient concentrations, microbial levels, microplastics, dissolved oxygen, phytoplankton concentrations, and fish densities have improved since the COVID-19 pandemic-related lockdown period. Discharges into waterways have reduced because of a decline in anthropogenic activities such as tourism, business and commercial interests, agriculture, and industry. Selvam et al. (2020) stated that the decrease in heavy metals and bacterial burdens was attributed to decreasing industrial and agricultural activity, respectively.

However, the decline in fishing activity has revealed a lack of fish and fishery products on the market, which has increased consumer burden and reduced income for inland fishermen. The price of catch in wholesale markets and landing hubs was on the rise as a result of the rising cost of fishing operations. The recommendations to mitigate the impacts of the COVID-19 pandemic are summarized in Tables 8.8 and 8.9.

Table 8.8 Recommendations to mitigate the impacts of the COVID-19 pandemic on inland capture fisheries

Impacts of COVID-19	Consequences	Recommendations
Increased number of people engaged in inland capture fishing activities	Upsurge in the awareness of the importance of inland fisheries as a safeguard for those who lost jobs in the cities Increase in the number of fishers who gained income from inland fishing activities	Raise the awareness of the existing fisheries laws and regulations of the local people who have engaged fishing practices at their hometowns
Decreased number of fishing vessels in operation	Shortage in the supply of fish and fishery products Recovery of fishery and natural resources due to decreased fishing pressure	Issue a special permit to fishers to sustain inland fishing activities
Increased price of catch in wholesale markets/landing centers	Additional burden for consumers Increase in the income for inland fishers	Provide incentives or discounts to consumers for them to afford to buy fish and fishery products
Increased cost of fishing operations	Additional burden for fishers	Provide financial support to fishers and enhance low interest loans

Table 8.9 Recommendations to mitigate the impacts of the COVID-19 pandemic on fisheries management

Impacts of COVID-19	Probable consequences	Recommendations
Cutting-edge technology to stop IUU Illegal, Unreported, and Unregulated) fishing	Increase in the number of apprehensions of IUU fishing activities	Increase the private sector's involvement, as well as that of pertinent international agencies and organizations, in the use of innovative, inexpensive technologies to stop IUU fishing
Frequency of implementation of port State measures	Increase in the number of occurrences of IUU fishing activities	Adopt remote Port State Measures (PSM) implementation by using online platforms
Implementation of Monitoring, Control and Surveillance (MCS)	Development in the oversight of fishing vessel operations	Increase the use of cutting-edge and economical technology to maintain the performance of MCS activities
Decrease of Monitoring, Control and Surveillance (MCS)	Interruption in the monitoring of the activities of fishing vessels	Intensification the use of cutting-edge and cost-effective technology to maintain the performance of MCS activities
Decrease of regularity to organize regular/routine data collection for fish stock assessment and monitoring of shared stocks/transboundary species	Gaps in data series for data base management	Explore innovative technologies and techniques to sustain the monitoring of transboundary stocks Enhance the collaboration with enumerators Utilize the available electronic platforms (e.g. e- logbooks, eACDS)
Decreased conduct of research/project activities	Decrease in the availability of information to support and sustain fisheries management	Investigate cutting-edge tools and methods when performing research Collective cooperation with enumerators (e.g. online communication) Intensification local communities' ability to conduct research activities (e.g. data collection)

8.8 Conclusion

Most of the world economy's sectors have been influenced by COVID-19. Communities and stakeholders in the fisheries and aquaculture sector have suffered because of the ongoing, restrictive measures imposed on travel, movement, and transportation. The current study shows that COVID-19 has caused several challenges for fishermen and other supply chain participants, including limited input supply, inadequate technical support, an inability to market their products, a lack of transportation to the market, export restrictions on fish and fishery products, and low fish prices in ASEAN nations, with a focus on Brunei Darussalam. Due to these issues, fish farmers could get disinterested in raising fish and related products. A widening imbalance between supply and demand in the fish food chain because of numerous COVID-19-related regulations has led to a downturn in the fishing industry and other stakeholders in the fish and aquaculture sectors.

Brunei Darussalam offers several benefits for potential investors in the goods and related industries of the aquaculture industry, due to its relatively open economy and advantageous geographic location in South-East Asia. The advantages of engaging in aquaculture in Brunei Darussalam include FDI incentives, a strategic position, infrastructure coverage, a high-quality environment, and a low sensitivity to natural disasters. However, there are risks and difficulties associated with investing in Brunei Darussalam's aquaculture. As a result, the country's aquaculture industry's numerous risks and problems are outlined, along with the relevant management techniques that can offer solutions to stop these risks and challenges from having an impact on the industry and its investment. Investors and industry participants should implement these steps even in the absence or absence of these problems to further enhance product quality and boost industry competitiveness. Since these policies affect everyone, including investors, business leaders, and the government of Brunei Darussalam, they must be pursued collaboratively. The aquaculture sector and its participants must be productive while resolving the difficulties the sector faces to become sustainable if they are to attract investment. According to the idea of "One Health," which links human health to that of animals, the environment, and human health, the wellbeing of all living things is crucial. A healthy source of aquatic animal protein can be produced using biosecurity procedures and good aquaculture methods.

Also, it is advised that investors follow certain best practices because doing so will benefit the entire sector. These include: (1) investing in small-scale businesses, which will ensure the industry's economic sustainability over time; (2) investing in research and development and technologies, which will help raise the productivity and quality of fish and seafood products; and (3) investing in spin-off industries, which will support the various activities of the aquaculture industry. To develop institutional capacity, it is crucial to collaborate with organizations and institutions like SEAFDEC and the relevant divisions under the Department of Fisheries in Brunei such as the Aquaculture Industry Division and the Aquatic Animal Health and Laboratory Services Centre. For foreign investors, strengthening institutional support especially from the Government of Brunei is specific to aquaculture is a crucial first step. In the near future, the economy of Brunei Darussalam in Asia will be strengthened by the implementation of the sectoral regulatory framework provided by the Government of Brunei, as well as new investment policies and incentives geared specifically to the aquaculture sector. In these contexts, this review paper provides recommendations for how to deal with the COVID-19 post-crisis and its effects on aquaculture and the seafood chain, focusing on the enhancement of capital source and financial support to those involved in the fisheries industry such as the possibility of offering interest-free loans and business-matching programme for those involved in the fisheries industry to assist them become more resilient. This review paper could be an effort for additional frontiers of actions into the major effects of COVID-19 on the aquaculture and fishing industries in Brunei Darussalam and other ASEAN nations.

Acknowledgements The authors are grateful to the report written by Mr. Ron Allan Estrebillo and Haruko Hiramoto under the Trade and Investment Cluster at the ASEAN-Japan Centre (AJC). The authors wish to express sincere gratitude to the all the staffs to the Department of Fisheries, Brunei Economic Development Board, Department of Statistics, Brunei Darussalam. The authors also indebted to the secretariat of SEAFDEC for their comprehensive report on ASEAN aquaculture ana Fisheries during Covid-19 Pandemic.

References

Alam GMM, Sarker MNI, Gatto MBH, Naziri D (2022) Impacts of COVID-19 on the fisheries and aquaculture sector in developing countries and ways forward. Sustainability 14(3):1071

Albastaki A, Naji M, Lootah R, Almeheiri R, Almulla H, Almarri I, Alreyami A, Aden A, Alghafri R (2021) First confirmed detection of SARS-COV-2 in untreated municipal and aircraft wastewater in Dubai, UAE: the use of wastewater-based epidemiology as an early warning tool to monitor the prevalence of COVID-19. Sci Total Environ 760:143350

Bandala ER, Kruger BR, Cesarino I, Leao AL, Wijesiri B, Goonetilleke A (2021) Impacts of COVID-19 pandemic on the wastewater pathway into surface water: a review. Sci Total Environ 774:145586

Bates AE, Primack RB, Moraga P, Duarte CM (2020) COVID-19 pandemic and associated lockdown as a "Global Human Confinement Experiment" to investigate biodiversity conservation. Biol Conserv 248

Beare D, Hölker F, Engelhard GH, McKenzie E, Reid DG (2010) An unintended experiment in fisheries science: a marine area protected by war results in Mexican waves in fish numbers-at-age. Naturwissenschaften 97(9):797–808

Bondad-Reantaso MG, Mackinnon B, Bin H, Jie H, Kathy T, Surachetpong W, Alday-Sanz V, Salman M, Edgar B, Iddya K, Larry H, Keith S, Manuel B, Alessandro L, Agus S, Nihad F, Rohana S, Árni MM, Mohamed S (2020) Viewpoint: SARS-CoV-2 (the cause of. COVID-19 in humans) is not known to infect aquatic food animals nor contaminate their products. Fisheries and Aquaculture Department, Food and Agriculture Organization of the United Nations, Rome, Italy. Asian Fish Sci 33:74–78

Bouey J (2020) From Sars to 2019-coronavirus (NCOV): US-China collaborations on pandemic response. Curr Polit Econ North West Asia 29(1):1–22

Braga F, Scarpa GM, Brando VE, Manfè G, Zaggia L (2020) COVID-19 lockdown measures reveal human impact on water transparency in the Venice Lagoon. Sci Total Environ 736:139612

Chun Z (2020) Coronavirus hits sustainable aquaculture. China Dialogue Ocean, Beijing

Cooke SJ, Twardek WM, Lynch AJ, Cow IG, Olden JD, Funge-Smith S, Lorenzen K, Arlinghaus R, Chen Y, Weyl OLF (2021) A global perspective on the influence of the COVID-19 pandemic on freshwater fish biodiversity. Biol Conserv 253:108932

Corlett RT, Primack RB, Devictor V, Maas B, Goswami VR, Bates AE, Koh LP, Regan TJ, Loyola R, Pakeman RJ, Cumming GS (2020) Impacts of the coronavirus pandemic on bio-diversity conservation. Biol Conserv 246

Degnarain N (2020) Six places where oceans, rivers and marine life have rebounded during the coronavirus pandemic. https://www.forbes.com/sites/nishandegnarain/2020/05/16/six-places-where-oceans-rivers-and-marine-life-have-rebounded-during-the-coronavirus-pandemic/?sh=43d597f63fb0

Dente SMR, Hashimoto S (2020) COVID-19: a pandemic with positive and negative outcomes on resource and waste flows and stocks. Resour Conserv Recycl 161

Department of Economic Planning and Statistics, Ministry of Finance and Economy (MOFE) (2020a) Gross domestic product by kind of economic activity at constant prices, 2010–2019. http://www.deps.gov.bn/SitePages/eData%20library.aspx

Department of Economic Planning and Statistics, Ministry of Finance and Economy (2020b) Gross domestic product growth rates by type of expenditure at constant prices, 2010–2019. http://www.deps.gov.bn/SitePages/eData%20library.aspx

Department of Economic Planning and Statistics, Ministry of Finance and Economy (MOFE) (2020c) Share in gross value added by kind of economic activity at current prices, 2010–2019. http://www.deps.gov.bn/SitePages/eData%20library.aspx

Department of Fisheries, Ministry of Primary Resources and Tourism (DOF, MPRT) (2021) DOF Strategic Plan for Investment Potential of Brunei Aquaculture Industry for Investors. Slides presented at online meeting of ASEAN-Japan Centre and Department of Fisheries, Brunei Darussalam

Early R, Bradley BA, Dukes JS, Lawler JJ, Olden JD, Blumenthal DM, Gonzalez P, Grosholz ED, Ibañez I, Miller LP, Sorte CJ (2016) Global threats from invasive alien species in the twenty-first century and national response capacities. Nat Commun 7(1):1–9

Edward JP, Jayanthi M, Malleshappa H, Jeyasanta KI, Laju RL, Patterson J, Raj KD, Mathews G, Marimuthu AS, Grimsditch G (2020) COVID-19 lockdown improved the health of coastal environment and enhanced the population of reef-fish. Mar Pollut Bull 165

Food and Agriculture Organization (FAO) (2020a) Summary of the impacts of the COVID-19 pandemic on the fisheries and aquaculture sector: addendum to the State of World Fisheries and Aquaculture 2020. FAO, Rome. https://doi.org/10.4060/ca9349en

Food and Agriculture Organization (FAO) (2020b) Best practices for developing surveys and questionnaires on the impacts of COVID-19 on fisheries and aquaculture. https://www.fao.org/fileadmin/user_upload/faoweb/FI/COVID19/Surveys_and_questionnairesCOVID.pdf

Food and Agriculture Organization (FAO) (2020c) The effect of COVID-19 on fisheries and aquaculture in Asia. FAO. https://doi.org/10.4060/ca9545en

Food and Agriculture Organization (FAO) (2020d) The State of World Fisheries and Aquaculture 2020. FAO. https://doi.org/10.4060/ca9229en

Food and Agriculture Organization of the United Nations (FAO) (2020e) Blockchain application in seafood traceability. Retrieved from http://www.fao.org/3/ca8945en/CA8945EN.pdf

Food and Agriculture Organization (FAO) (2021) The impact of COVID-19 on fisheries and aquaculture food systems, possible responses. FAO. https://doi.org/10.4060/cb2537en

Food and Agriculture Organization (FAO) (2022) Q&A: COVID-19 pandemic—impact on fisheries and aquaculture. FAO. Food and Agriculture Organization of the United Nations. https://www.fao.org/2019-ncov/q-and-a/impact-on-fisheries-and-aquaculture/en

Ferrer AJG, Pomeroy R, Akester MJ, Muawanah U, Chumchuen W, Lee WC, Hai PG, Viswanathan KK (2021) COVID-19 and small-scale fisheries in Southeast Asia: impacts and responses. Asian Fish Sci 34:99–113

Islam MM, Khan MI, Barman A (2021) Impact of novel coronavirus pandemic on aquaculture and fisheries in developing countries and sustainable recovery plans: case of Bangladesh. Mar Policy 131:104611

Jalal KCA (2020) Mitigation strategies for the Covid-19 impacts on aquaculture in Malaysia. International Forum on Aquaculture for Silk Road Countries. Jointly organized by Food and Agriculture Organization of the United Nations, Network of Aquaculture Centers in Asia-Pacific Chinese Academy of Fishery Sciences Qingdao, China, 24–25th Nov 2020

Jalal KCA, John A, Abdullah R, Musa N, Ashraf MA (2021) Impact of COVID-19 on aquaculture and fisheries in ASEAN countries: some aspects of challenges, mitigations for future strategies in Malaysia. Desalin Water Treat 241:331–339

Kaewnuratchadasorn P, Smithrithee M, Sato A, Wanchana W, Tongdee N, Sulit V (2020) Capturing the impacts of COVID-19 on the fisheries value chain of Southeast Asia. 18(2):2–8. http://repository.seafdec.org/handle/20.500.12066/6557

Lala-Pritchard T, O'Leary M, Lim F, Huso D (2020) World Fish 2020 Annual Report

Liu D, Thompson JR, Carducci A, Bi X (2020) Potential secondary transmission of SARS-CoV-2 via wastewater. Sci Total Environ 749:142358

Lotliker AA, Baliarsingh SK, Shesu RV, Samanta A, Naik RC, Nair TB (2021) Did the coronavirus disease 2019 lockdown phase influence coastal water quality parameters off major Indian cities and river basins? Front Mar Sci

Najah A, Teo FY, Chow MF, Huang YF, Latif SD, Abdullah S, Ismail M, El-Shafie A (2021) Surface water quality status and prediction during movement control operation order under COVID-19 pandemic: case studies in Malaysia. Int J Sci Environ Technol 18:1009–1018

Nelson K, Surachetpong W (2020) Viewpoint: SARS-CoV-2 (the cause of COVID-19 in humans) is not known to infect aquatic food animals nor contaminate their products. Asian Fish Sci. https://doi.org/10.33997/j.afs.2020.33.1.009

Nesar A, Azra MN (2022) Aquaculture production and value chains in the COVID-19 pandemic. Curr Environ Health Rep 9:423–435. https://doi.org/10.1007/s40572-022-00364-6

Organization for Economic Co-operation and Development (OECD) (2021) COVID-19 and multilateral fisheries management. https://www.oecd.org/coronavirus/policy-responses/covid-19-and-multilateral-fisheries-management-cc1214fe/#boxsection-d1e39

Praveena SM, Aris AZ (2021) The impacts of COVID-19 on the environmental sustainability: a perspective from the Southeast Asian region. Environ Sci Pollut Res 1–8

Ritchie H, Mathieu E, Rodés-Guirao L, Appel C, Giattino C, Ortiz-Ospina E, Hasell J, Macdonald B, Beltekian D, Roser M (2021) Coronavirus Pandemic (COVID-19). https://ourworldindata.org/coronavirus

Ron AE, Haruko H (2021) Brunei Darussalam aquaculture feasibility study for investment. ASEAN-Japan Centre (AJC), p 116

Sardain A, Sardain E, Leung B (2019) Global forecasts of shipping traffic and biological invasions to 2050. Nat Sustain 2(4):274–282

Seitzinger SP, Mayorga E, Bouwman AF, Kroeze C, Beusen AH, Billen G, Van Drecht G, Dumont E, Fekete BM, Garnier J, Harrison JA (2020) Global River nutrient export: a scenario analysis of past and future trends. Glob Biogeochem Cycles 24:40

Selvam S, Jesuraja K, Venkatramanan S, Chung SY, Roy PD, Muthukumar P, Kumar M (2020) Imprints of pandemic lockdown on subsurface water quality in the coastal industrial city of Tuticorin, south India: a revival perspective. Sci Total Environ 738:139848

Sharma RS, Panthari D, Semwal S, Uniyal T (2021) The impact of the COVID-19 pandemic on green societies. Springer, Cham. Aftermath of industrial pollution, post COVID-19 quarantine on environment, pp 141–167

Southeast Asian Fisheries Development Center (SEAFDEC) (2020a) Report of the SEAFDEC Webinar on Impact of the COVID-19 on Fisheries and Aquaculture, in Southeast Asia July 2020. SEAFDEC, p 100

Southeast Asian Fisheries Development Center (SEAFDEC) (2020b) Study on the impacts of COVID-19 pandemic on the fisheries sector of the ASEAN-SEAFDEC member countries. The Regional Workshop for the Finalization of the Study Report on Impacts of COVID-19 Pandemic on the Fisheries Sector of the ASEAN-SEAFDEC Member Countries, 20 Apr 2022, p 84

Southeast Asian Fisheries Development Center (SEAFDEC) (2021) Fishery statistical bulletin of Southeast Asia 2018. Southeast Asian Fisheries Development Center. http://repository.seafdec.org/handle/20.500.12066/6601

Southeast Asian Fisheries Development Center (SEAFDEC) (2022) Fisheries Country Profile: Brunei Darussalam. Published 20th Mar 2023. http://www.seafdec.org/fisheries-country-profile-brunei-darussalam

Sun X, Liu J, Wang J, Tian L, Zhou Q, Li J (2020) Integrated monitoring of lakes turbidity in Wuhan, China during the COVID-19 epidemic using multi-sensor satellite observations. Int J Digit Earth 14:443–463

World Health Organization (WHO) (2020a) Q&A on Coronaviruses (COVID-19). WHO, Geneva, Switzerland. Available online: https://www.who.int/emergencies/diseases/novel-coronavirus-2019/question-and-answers-hub/q-a-detail/q-a-coronaviruses. Accessed on 29 July 2021

World Health Organization (WHO) (2020b) Infection prevention and control during health care when novel coronavirus (nCoV) infection is suspected: interim guidance, 25 Jan 2020 (No. WHO/2019-nCoV/IPC/2020.2)

World Health Organization (WHO) (2021) Archived: WHO Timeline—COVID-19. https://www.who.int/news/item/27-04-2020-who-timeline---covid-19

Xu Z, Shi L, Wang Y, Zhang J, Huang L, Zhang C, Liu S, Zhao P, Liu H, Zhu L, Tai Y (2020) Pathological findings of COVID-19 associated with acute respiratory distress syndrome. Lancet Respir Med 8(4):420–422

Yunus AP, Masago Y, Hijioka Y (2020) COVID-19 and surface water quality: improved Lake water quality during the lockdown. Sci Total Environ 731:139012

Yusoff FM, Abdullah AF, Aris AZ, Umi WAD (2021) Impacts of COVID-19 on the aquatic environment and implications on aquatic food production. Sustainability 13:11281

Chapter 9
Fishery Resilience in Covid-19 Post-Pandemic Era: Sustainable Environmental Approach for Potential Challenges in Asia

Akbar John, Ahmed Jalal Khan Chowdhury, Mohammad Bodrul Munir, Rose Abdullah, Zaima Azira Zainal Abidin, and Noor Faizul Hadry Nordin

Abstract The COVID-19 pandemic has had a substantial impact on the global fishery industry, causing widespread disruption to supply chains and market demand. This has led to significant economic challenges for fishing communities, particularly in developing countries where fishing is a major source of livelihood. In the challenge of the pandemic, it is important to consider ways to build resilience in the fishery industry to ensure its continued viability in the time to come. One approach to achieving fishery resilience is to focus on sustainable fishing practices. This involves implementing measures to minimize the impacts of fishing on the natural aquatic environment by reducing human intervention such as bycatch and protecting critical habitats. Additionally, investment in aquaculture can help to reduce dependence on wild-capture fishing and provide a more controlled and sustainable source of fish

---

A. John (✉)
College of Marine Sciences and Aquatic Biology, University of Khorfakkan, Sharjah, United Arab Emirates
e-mail: akbarjohn50@gmail.com

A. J. Khan Chowdhury · M. B. Munir · R. Abdullah
Faculty of Agriculture, Universiti Islam Sultan Sharif Ali (UNISSA), KM 33, Jalan Tutong, Kampung Sinaut, Tutong TB1741, Brunei Darussalam
e-mail: ahmed.chowdhury@unissa.edu.bn

R. Abdullah
e-mail: rose.abdullah@unissa.edu.bn

Z. A. Zainal Abidin
Department of Biotechnology, Kulliyyah of Science, International Islamic University Malaysia, 25200 Kuantan, Pahang Darul Makmur, Malaysia
e-mail: zzaima@iium.edu.my

N. F. H. Nordin
International Institute for Halal Research and Training, International Islamic University Malaysia, Level 3, KICT Building, 53100 Jalan Gombak, Selangor, Malaysia
e-mail: faizul@iium.edu.my

D. Jones and A. J. Khan Chowdhury (eds.), *Environmental Resilience in the Pandemic Years 2020–2021*, Environmental Science and Engineering,
https://doi.org/10.1007/978-3-031-47757-7_9

for human consumption. Another important aspect of building fishery resilience is improving supply chain management. This involves establishing more efficient and effective processes for catching, processing, and distributing fish and seafood products, to minimize waste and ensure that products reach markets in a timely manner. Improving supply chain management can also help to reduce the economic impacts of the pandemic by ensuring that fish and seafood products are available for sale even in the face of market disruptions. In conclusion, the COVID-19 pandemic has had a profound impact on the global fishery industries especially Asia–Pacific region, highlighting the need for greater resilience in this critical sector. By means of focusing on sustainable fishing practices and improving supply viability for generations to come chain management, the industry can develop a more sustainable and resilient future, ensuring its viability for generations into the near future.

Keywords Sustainable practice · Fishery sector · Supply chain · Fishery management · COVID-19

9.1 Introduction

9.1.1 Effects of COVID-19 on Global Fisheries and Aquaculture Production

Global changes in consumer demand, market disruption, and logistical challenges in implementing strict containment measures, such as lockdowns, curfews, physical separation in operations and onboard vessels, and port restrictions, have all contributed to the COVID-19 pandemic's profound effects on fisheries and aquaculture. The prices of fisheries and aquaculture products in Asia and the Pacific fell because of these lockdowns, which decreased demand. Numerous aquaculture or fishing fleets stopped operating or scaled back their operations as a result of their labor being unprofitable, particularly during the pandemic waves of 2020 (FAO 2021).

Similarly, surveys at sea frequently came to an end, endangering the gathering of vital information for the assessment of stocks over time and place. In additional occurrences, it was difficult to ensure hygienic procedures (such as physical separation between crew members at sea) or there were insufficient supplies, which prevented the deployment of scientific observers at sea (e.g., face masks and gloves). The data gathering from aquaculture facilities was also negatively impacted. In several nations, the customary data collecting on fisheries and aquaculture at landing locations has been frequently stopped. These challenges were not homogeneous among countries or even within the same country, as some had better institutional, financial, technological, and digital capacities to develop solutions. Overall, COVID-19 brought a new set of challenges to national fisheries statistics systems and operations in these regions (FAO 2022).

9.1.2 *Global Market and Demand on Fishery Sector*

The demand for fishery products has been growing globally due to the increasing population and rising disposable incomes, leading to greater consumption of seafood. According to the Food and Agriculture Organization (FAO) of the United Nations, the global production of fish and fishery products has increased from 131 million tonnes in 2000 to 179 million tonnes in 2019 (FAO 2020a). This growth in production can be attributed to the increasing demand for fish and seafood as a source of protein, as well as advances in aquaculture technology, which has allowed to produce fish in controlled environment. In terms of regional demand, Asia–Pacific region accounts for most of the global demand for fishery products, driven by the large and growing populations in countries such as China and India. In 2019, Asia accounted for over 60% of the global demand for fishery products, with Europe and North America accounting for the remaining share (FAO 2020b).

The global market for fishery products is expected to reach USD 372.7 billion by 2026, growing at a CAGR of 4.5% from 2021 to 2026. This growth is expected to be driven by factors such as increasing demand for fish as a source of protein, as well as population growth and urbanization. Additionally, the increasing popularity of health and wellness trends and the recognition of the health benefits associated with consuming fish is also expected to drive growth in the fishery sector (Mordor Intelligence 2021). Notwithstanding some preliminary statistics it is still too early to quantify all of the effects of the lockdown's decline in fisheries (Ortega et al. 2020).

Global fisheries provided 96.4 million tons of fish to society prior to the pandemic, 88% of which were used for direct human consumption before the Pandemic Covid-19. The primary fishing industry employs 38.89 million people worldwide, and there are 4.56 million active fishing vessels. This industry exports USD 164.1 billion a year (FAO 2020c, 2022) (Table 9.1).

In 2020, FAO estimated 60 million tonnes live weight of aquatic products exported globally, excluding algae, for a total of USD 151 billion. Compared to the record high of 67 million tonnes, valued at USD 165 billion, attained in 2018, this indicates a significant fall (8.4% in value and 10.5% in volume).

Global fisheries and aquaculture exports' total value increased at an average annual growth rate of 6.9% in nominal terms and 3.9% in real terms (adjusted for inflation) from 1976 to 2020, while their quantity increased at an average annual growth rate of 2.9% during the same period (Fig. 9.1) (FAO 2020d, 2022).

9.2 Global Marine Fisheries Ecosystem

The growth of the fishery sector may be impacted by various challenges such as over-fishing, illegal fishing, and sustainability concerns. Overfishing is a major concern as it can lead to a decline in the number of fish species and a decrease in the overall size of fish populations. This, in turn, can lead to a reduction in the availability of fish and

Table 9.1 World fisheries and aquaculture production, utilization and trade

	1990s	2000s	2010s	2018	2019	2020
	Average per year					
	Million tonnes (live weight equivalent)					
Production						
Capture:						
Inland	7.1	9.3	11.3	12.0	12.1	11.5
Marine	81.9	81.6	79.8	84.5	80.1	78.8
Total capture	**88.9**	**90.9**	**91.0**	**96.5**	**92.2**	**90.3**
Aquaculture:						
Inland	12.6	25.6	44.7	51.6	53.3	54.4
Marine	9.2	17.9	26.8	30.9	31.9	33.1
Total aquaculture	**21.8**	**43.4**	**71.5**	**82.5**	**85.2**	**87.5**
Total world fisheries and aquaculture	**110.7**	**134.3**	**162.6**	**178.9**	**177.4**	**177.8**
Utilization[2]						
Human consumption	81.6	109.3	143.2	156.8	158.1	157.4
Non-food uses	29.1	25.0	19.3	22.2	19.3	20.4
Population (billions)[3]	5.7	6.5	7.3	7.6	7.7	7.8
Per capita apparent consumption (kg)	14.3	16.8	19.5	20.5	20.5	20.2
Trade						
Exports—in quantity	39.6	51.6	61.4	66.8	66.6	59.8
Share of exports in total production (%)	35.8	38.5	37.7	37.3	37.5	33.7
Exports—in value (USD 1 billion)	46.6	76.4	141.8	165.3	161.8	150.5

Source FAO (2022); https://population.un.org/wpp

a rise in prices, potentially reducing the demand for fishery products. Illegal fishing is another challenge facing the fishery sector. This refers to fishing that is carried out in violation of national and international fishing regulations, often in marine protected areas or in areas where fishing is prohibited. Illegal fishing can lead to a decline in fish populations and damage to marine ecosystems, reducing the potential for sustainable fishing and undermining the long-term viability of the fishery sector (FAO 2018).

Fisheries functioning and their effects on marine resources and ecosystems can be understood through disruptions in local and global value chains of seafood production. With more knowledge, market diversification strategies can be promoted as a crucial safeguard for the resilience and sustainability of fisheries in the future (Knight et al. 2020). Smith et al. (2020) reported that the declining fishing pressure associated with lockdown periods resulted in increased stocks for some species, but the landing of most commercial species remained the same, probably due to the relatively short lockdown period. In fact, reduced fishing pressure had positive effects only on the fast-growing species, such as deep-water shrimp.

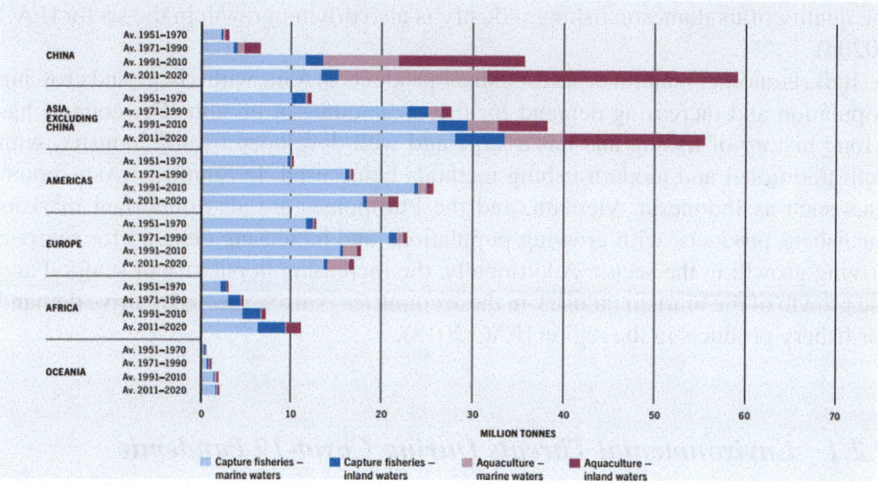

Fig. 9.1 World capture fisheries and aquaculture production. *Source* FAO (2022); https://population.un.org/wpp

However, those positive effects disappeared with the relaxation of the lockdown (Coll et al. 2021). Thus, short-term increases of biomass for fast-growing, small-sized organisms during the COVID-19 lockdown period quickly vanished when fishing resumed, indicating that longer periods of non-intervention are needed to allow for ecosystems' complete recovery. Overall, marine fisheries ecosystem changed dramatically during COVID-19. In fact, global industrial fishing activity was reduced by 6.5% at the end of April 2020 compared to same period in 2019, or by 10% if calculated from the date the global pandemic was declared. Regionally, fishing activity reductions varied (FAO 2020a).

Sustainability is also a growing concern in the fishery sector, as the increasing demand for fish and seafood products is putting pressure on marine ecosystems and leading to overfishing and the depletion of fish stocks. To address these sustainability concerns, many countries and organizations are implementing sustainable fishing practices and promoting responsible fishing methods to ensure that fish stocks are managed in a responsible and sustainable manner. The demand for fishery products is expected to continue to grow in the coming years, driven by factors such as increased demand for fish as a source of protein, as well as population growth and urbanization (FAO 2020a).

Nevertheless, Asia is the largest market for fishery products globally, accounting for over 60% of the global demand. The growing population, rising disposable incomes, and increasing consumption of seafood are driving the demand for fishery products in Asia. China is the largest market for fishery products in Asia, accounting for most of the regional demand. The country has a large and growing population, and the increasing disposable incomes and changing dietary habits are leading to greater consumption of seafood. In addition, the Chinese government's focus on improving the quality of its domestic fishing industry is also driving growth in the sector (FAO 2020d).

India is another major market for fishery products in Asia, with a large and growing population and increasing demand for fish as a source of protein. The country has a long history of fishing and has a large and well-developed fishing industry, with both traditional and modern fishing methods being used. In Southeast Asia, countries such as Indonesia, Vietnam, and the Philippines are also important markets for fishery products, with growing populations and increasing demand for seafood driving growth in the sector. Additionally, the increasing popularity of seafood and the growth of the tourism industry in these countries is also expected to drive demand for fishery products in the region (FAO 2018).

9.2.1 Environmental Threats During Covid-19 Pandemic

Environmental factors that are biophysical, managed, and built constitute potential causes of harm to human life, health, income, and property from ecological risks. Both direct and indirect exposure with COVID-19 can result in the infection of respiratory viruses. WHO has encouraged the public to wear personal protective equipment like gloves and face masks in order to control and prevent the spread of this dangerous illness. There is also a dearth of advice on how to properly discard discarded gloves and face masks. Inadequate disposal facilities for biohazardous items make the situation worse, and thus poses a serious concern to both the marine environment and human health as a new type of marine contaminant that could contaminate water supplies (WHO 2020). It has raised the danger of plastic pollution in the ocean's aquatic ecology. Inadequate disposal facilities for biohazardous items make the problem worse, and thus poses a serious concern to both human health and the marine environment as a new type of marine pollutant that may contaminate aquatic goods. Also inevitable is the rise in single-use plastics during the COVID-19 pandemic. Plastics are used in both takeout food containers and frontline soldiers' protective gear. To prevent cross-contamination, several restaurants and grocery stores now forbid patrons from bringing their own containers or shopping bags. Long-term exiting of these situations could increase the risk of plastic pollution in aquatic environments (Chuan et al. 2021). Some aspects of the risk assessment of COVID-19 on aquatic environment are reviewed in Table 9.2 (Yusoff et al. 2021).

Table 9.2 Risk probability of COVID-19 on aquatic environment (Ref: Yusoff et al. 2021)

Sectors	Environmental threats	Risk	Risk mitigation	Reference
Fisheries	Disposal of Medical Waste	Paucity of evidence on accurately dispose the used face masks and gloves might negative impact on the marine environment	Awareness of Pragmatic policies and governing agendas	Mejjad et al. (2021)
Fisheries	Long term impact of chemicals exposure	Long-term impact on human exposure and aquatic ecosystem due to excessive and erratic use of chemical disinfectants globally used	Restriction on releases of chemical disinfectants for the reduction of long-term impact	Ghafoor et al. (2021)
Fisheries	Dumping of Biohazards materials	Inadequate disposal facilities to utilize the biodiesel with the biohazard materials that have negative impact on the marine environment as marine pollutant	Pragmatic policies and governing agendas management practices, good manufacturing practices (GMP)	Sarkodie and Owusu (2021)
Fisheries	Utilization of non-biodegradable materials	The increase of single use plastics during the COVID-19 pandemic will increase the risk in oceans and estuaries marine life	Continuous monitoring and restriction on the utilization of non-biodegradable plastics in the aquatic environment	Silva et al. (2021)

9.3 Economic Challenges on Fishing Communities and Marine Environment

Fishing communities in Asia face a variety of economic challenges that affect their livelihoods and their ability to make a living from fishing. Some of the key economic challenges facing fishing communities in Asia include:

Overfishing

Overfishing has reduced the amount of fish available for harvest, which in turn has driven up the cost of fish and reduced the profitability of fishing. This has had a particularly negative impact on small-scale fishing communities who are unable to compete with larger fishing operations Knapp and Roheim (2007).

Competition from Aquaculture

The growth of aquaculture in Asia has increased competition for fish and seafood products, reducing the demand for wild-caught fish and negatively impacting the livelihoods of fishing communities (Song and Chuenpagdee 2014).

Access to markets

Fishing communities in remote areas often face challenges in accessing markets, which limits their ability to sell their catch and earn a livelihood. This can be due to a lack of infrastructure, high transportation costs, and limited access to information and technology (Daw et al. 2006).

9.4 Depletion of Fishing Stocks

The depletion of fishing stocks has reduced the number of fish available for harvest, which has negatively impacted the livelihoods of fishing communities. In some cases, fishing communities have had to turn to alternative livelihoods, such as agriculture or tourism, to make a living (Williams and Kittinger 2017).

Climate change

Climate change is having a significant impact on the fishing industry in Asia, affecting fishing patterns, migration patterns of fish, and the productivity of fishing grounds. This has made it more difficult for fishing communities to make a living from fishing and has increased the need for alternative livelihood options (FAO 2016).

9.5 Fishery Resilience and Sustainable Fishery

Sustainable fishery practices can help increase the resilience of the fishery sector in the future by reducing the negative impacts of fishing on the environment and fishing stocks, and by ensuring the long-term viability of the fishing industry.

Maintaining healthy fish stocks

By using sustainable fishing methods that minimize bycatch and reduce the impact of fishing on the marine environment, it is possible to maintain healthy fish stocks and preserve the long-term viability of fishing grounds. This helps to ensure that fishing communities have a reliable source of fish for future generations (Ota et al. 2020).

Protecting the marine environment

Sustainable fishing practices help to protect the marine environment by reducing the impact of fishing on the ecosystem. This includes measures such as reducing bycatch, reducing the use of harmful fishing gear, and protecting sensitive marine habitats (Gephart et al. 2021).

Improving profitability

By reducing the negative impact of fishing on the environment and fish stocks, sustainable fishing practices can help to improve the profitability of the fishing industry in the long term. This is because healthy fish stocks and a protected marine environment are essential for the sustainable development of the fishing industry (Pilling and Guénette 2019).

Encouraging responsible fishing

Sustainable fishing practices encourage fishing communities to adopt responsible fishing practices that protect the environment and fishing stocks. This helps to promote a culture of sustainability within the fishing industry and can help to ensure the long-term viability of the industry (Salas et al. 2020).

Improving food security

Sustainable fishing practices can help to improve food security by ensuring that there is a reliable supply of fish for future generations. This is particularly important in developing countries where fish is a critical source of protein for many people. Sustainable fishing practices play a crucial role in increasing the resilience of the fishery sector in the future. By reducing the negative impacts of fishing on the environment and fish stocks, and by promoting responsible fishing practices, it is possible to ensure the long-term viability of the fishing industry and to improve food security for future generations (Cashion et al. 2017). The effects of COVID-19 on the fisheries and aquaculture sector are multifarious. The present study has identified the key affected domains of the fisheries sector, these are the stakeholders, freshwater aquaculture, brackish water aquaculture, river and naturally sourced fisheries, offshore fisheries, and industry. Fishermen, farmers, auctioneers, and traders are the main stakeholders. The major impacts felt at the stakeholder level are the limited access to livelihood capital, disruption of strategies for securing a livelihood, increased vulnerability of livelihoods, and increase in food and nutritional insecurity (Mandal et al. 2021; Workie et al. 2021).

Due to market closures, rising freight costs, canceled flights, and border restrictions, unsold output raised the expense of feeding aquatic animals and their mortality rate in aquaculture. The production of fish and aquaculture reliant on export markets was more negatively impacted than that supplying domestic markets. However, the closure of the food service industries also had a significant impact on the home fresh fish and shellfish supplies. The impact on the countries in the Asia and Pacific region varied, with several reporting significant declines in aquaculture output and catch during the early weeks and months of the crisis, followed by increases as the industry responded as described in Table 9.3 (Alam et al. 2022).

Table 9.3 Common challenges in aquaculture and fisheries sector during Covid-19 pandemic

Effected areas	Common challenges	References
Fish farming practices	Difficulty of maintaining proper timing for fish farming	Manlosa et al. [35]
Fish and shrimp feed	Lack of demand for fish and shrimp fry by inadequate numbers of fish feed suppliers	Waiho et al. (2020) Sunny et al. (2021) Wiradana et al. (2021)
Unavailability of farming inputs	Inability to maintain daily farming expenses due the expensive operation cost of fish farm	Islam et al. (2021)
Overstocking of fishes	Inadequate number of buyers in the market caused decline of the revenue of the farm due to unsold marketable fish stocked in the farms	Zorriehzahra et al. (2020)
Fish mortality	Improper maintenance of the farms due to Covid restriction, Fish Disease was occurred, and fish mortality were spreading around the whole farms	Kumaran et al. (2021)
Shortage of labor	The unwillingness of labor force due to fear of a COVID-19 infection	Plagányi et al. (2021) Mandal et al. (2021)
Hatchery operation	Low demand for fish fry, fingerlings, and the and shrimp post larval price from the hatchery products	Hamzah and Nurdin (2021)
Transportation of fish and shrimp fry	Expensive transportation due to Covid-19 restrictions	Islam et al. (2021) Belton et al. (2021)

9.6 Aquaculture Investment and Impact on Fishery

The global aquaculture industry has experienced significant growth and investment over the past 30 years. According to the Food and Agriculture Organization of the United Nations (FAO), global aquaculture production increased from around 10 million tonnes in 1990 to over 110 million tonnes in 2019. Investment in aquaculture has been a key driver of this growth, as the industry has become increasingly focused on improving production methods, increasing efficiency, and reducing costs. The rise of new technologies and innovations in the sector, such as the development of new feed ingredients, improved breeding techniques, and the use of recirculating aquaculture systems, has also contributed to the growth of the industry (FAO 2020a).

In terms of regional investment, Asia has been the largest investor in the aquaculture industry, accounting for over 80% of global aquaculture production. The largest producers of aquaculture products in Asia are China, India, Indonesia, and Vietnam.

In recent years, there has also been significant growth in aquaculture production in Africa and South America, as these regions seek to meet the increasing demand for fish and seafood products. In terms of specific species, the top five species produced through aquaculture are carp, tilapia, salmon, pangasius, and shrimp. The production of these species has increased significantly in recent years, driven by the increasing demand for these products in the global market. The growth of the aquaculture industry over the past 30 years has been driven by significant investment in the sector, as well as advances in technology and the increasing demand for fish and seafood products. These trends are expected to continue in the coming years, as the industry continues to grow and evolve in response to changing market conditions and consumer demand (FAO 2020b).

The COVID-19 pandemic has had a significant impact on the global aquaculture industry, leading to increased investment in the sector as the demand for seafood products has increased. In the early months of the pandemic, the global food supply chain was disrupted, leading to supply shortages and increased demand for seafood products. This increased demand, combined with the reduced availability of wild-caught fish due to supply chain disruptions and reductions in fishing effort, led to higher prices for seafood products. As a result of these market conditions, many investors have seen the aquaculture sector as a more attractive investment opportunity. According to recent reports, investment in the aquaculture sector has increased in last five years, particularly in areas such as technology, research and development, and new product development. Nevertheless, some concepts were provided on this global snapshot about key challenges in terms of supply-driven input- and demand-driven out related marketing constraints due to the impacts of COVID-19 on aquaculture production and value chains (Nesar and Azra 2022). Additionally, the COVID-19 pandemic has also highlighted the importance of food security and self-sufficiency, leading to increased government investment in the aquaculture sector in many countries. This includes investments in research and development, infrastructure, and support for the development of new and existing aquaculture projects. It is difficult to provide specific statistical values for the increase in investment in the aquaculture sector after the COVID-19 pandemic, as data availability and methodology can vary by country and region. Bu the trend towards increased investment will continue as the demand for seafood products remains strong and the industry continues to grow and evolve (Jalal et al. 2021; FAO 2020e).

9.7 Supply Chain Management in Fishery Sector

Improving supply chain management is a key factor in building fishery resilience during and after the COVID-19 pandemic. The pandemic has highlighted the importance of having a resilient and effective supply chain in the fishery sector, as disruptions to the supply chain can lead to significant economic losses for fishers and fishing communities stated in Table 9.4 (Hossain and Alam 2021; Miao et al. 2022; Alam et al. 2022).

Table 9.4 Summary of the impacts of COVID-19 on the aquatic food supply chain (Ref: Alam et al. 2022)

Major domains of supply Chain	Impacts of COVID-19	References
Fishing	• Restricted access to capture Fisheries • Decreased duration of catching Time • Increased labor cost • Travel restrictions • Increased cost of inputs and Transportation • Undesired seed stock	Fiorella et al. (2021) Campbell et al. (2021) Ruiz-Salmón et al. (2021) Paradis et al. (2021) Cooke et al. (2021) Manlosa et al. (2021)
Processors	• Drop of rates of sale and Price • High-priced inputs • Controlled transportation due to restrictions	White et al. (2021) Bennett et al. (2020) Kumaran et al. (2021)
Cold Storage Facilities	• Fewer access to cold storage Facilities • Unanticipated loss due to the delicate nature of the product	Fahlevi et al. (2021) Kaewnuratchadasorn et al. (2020)

There are several ways in which supply chain management can be improved to build fishery resilience. (Yilmaz and Erdogan 2021). These include:

Improving traceability: Having a robust traceability system can help to ensure that products are safe and of high quality, while also reducing the risk of fraud and illegal activities in the supply chain. This is particularly important in the context of the COVID-19 pandemic, as concerns about food safety have increased.

Enhancing collaboration and communication: Effective collaboration and communication between actors in the supply chain is critical to ensuring that products are delivered to market in a timely and efficient manner. Improving communication and collaboration can also help to reduce the risk of supply chain disruptions.

Strengthening infrastructure: Investing in infrastructure, such as cold storage facilities, transportation systems, and processing plants, can help to ensure that products are delivered to market in good condition, reducing the risk of spoilage and waste.

Encouraging investment in technology: The use of technology, such as blockchain and digital traceability systems, can help to improve the efficiency and transparency of supply chain management. This can help to reduce the risk of fraud and improve the ability of actors in the supply chain to respond to disruptions.

Supporting small-scale fishers: Small-scale fishers often face significant challenges in accessing markets and getting their products to market in a timely and efficient manner. Supporting these fishers through initiatives such as training, technical assistance, and access to financing can help to build their resilience and ensure that their products reach market in good condition.

9.8 Chain of Aquatic Food Supply Disruption

The network of markets that makes up the aquatic food supply includes local, regional, and global markets. Transporting fish and fish products from the source to the final customer involves a lot of processes. The technology employed for this purpose vary considerably over the world, from conventional to highly industrial. Across the supply chain, COVID-19's impacts have had an impact on every activity. The important points at which COVID-19 impacts the primary, secondary, retailing, and export markets are indicated inside by the bold arrows on the left.

9.8.1 Challenges on Opportunities in Fishery Sector Towards Next Three Decades

The growth of the fishery sector over the next 30 years is expected to be influenced by a range of factors, including technological advancements, changes in consumer preferences, and the impact of pandemic diseases such as COVID-19. On the one hand, the development of new technologies and methods for sustainable fishing and aquaculture is expected to drive growth in the fishery sector. This includes the use of more efficient and sustainable fishing gear, the expansion of aquaculture production, and the development of new and innovative seafood products (De Silva 2020).

However, the impact of pandemic diseases such as COVID-19 is likely to present significant challenges to the growth of the fishery sector. The COVID-19 pandemic has already caused disruptions to global trade and the food supply chain, leading to supply shortages and increased prices for seafood products. Additionally, the risk of future pandemic diseases, such as new strains of influenza, is a concern for the fishery sector. Outbreaks of pandemic diseases can lead to reduced demand for seafood products, as consumers become concerned about the safety of seafood and the health risks associated with consuming it. To mitigate the impact of pandemic diseases on the fishery sector, it is important to invest in measures such as disease surveillance and control, food safety and traceability systems, and resilience planning. By improving the ability of the sector to respond to outbreaks of pandemic diseases, the risk of disruptions to the food supply chain can be reduced, helping to ensure the continued growth and stability of the fishery sector (Sveinsson 2021).

9.9 Importance of Blockchain Technology in Fishery Sector During Post Pandemic Era

Blockchain technology has the potential to revolutionize the fishery sector by improving transparency, traceability, and accountability throughout the supply chain. In the post-pandemic era, where there is an increasing demand for safe and sustainable seafood, blockchain technology can provide a solution to some of the challenges facing the fishery sector (Lee et al. 2019). One of the primary challenges in the fishery sector is the lack of transparency and traceability throughout the supply chain. Fish are often caught in one location, processed in another, and sold in a third location, making it difficult to track the origin of the fish and ensure that they have been caught and processed sustainably. Blockchain technology can provide a solution to this challenge by creating a decentralized and transparent system for tracking fish from the moment they are caught to the point of sale. Blockchain technology can also improve the accountability of actors in the fishery sector. By creating an immutable record of transactions, blockchain technology can help prevent fraud and ensure that fish are caught and processed legally and sustainably. Additionally, blockchain technology can help prevent human rights abuses in the fishery sector by ensuring that fish are not caught using forced labor or other unethical practices (FAO 2020c).

Furthermore, blockchain technology can enable consumers to make more informed choices about the seafood they consume. By providing information about the origin, species, and sustainability of the fish, blockchain technology can empower consumers to make more sustainable and ethical choices. This can also help incentivize fishery operators to adopt more sustainable practices, as they will be rewarded for their efforts through increased demand for their products. Several blockchain-based initiatives have already been implemented in the fishery sector. For example, the Fish coin project uses blockchain technology to create a traceability system for small-scale fishers in developing countries (Yilmaz and Erdogan 2021). The project provides fishers with mobile devices that allow them to record their catch, which is then tracked using blockchain technology. This creates a transparent record of the fish's journey from the moment it is caught to the point of sale, ensuring that consumers can make informed choices about the seafood they consume. Another example is the Seafood Chain project, which uses blockchain technology to create a traceability system for seafood products. The project tracks the origin of the fish and provides information about the sustainability of the fishery, allowing consumers to make more informed choices about the seafood they consume (De Silva and Karunarathne 2021).

9.10 Conclusion and Recommendations

Overall, the demand for fishery products in Asia is expected to continue to grow in the coming years, driven by the large and growing populations, rising disposable incomes, and increasing consumption of seafood. However, the growth of the fishery sector in Asia may be impacted by various challenges such as overfishing, illegal fishing, and sustainability concerns, and addressing these challenges will be crucial for the continued growth and development of the fishery sector in the region.

Fishing communities in Asia face a range of economic challenges that impact their livelihoods and their ability to make a living from fishing. Addressing these challenges will be crucial for the sustainable development of the fishing industry in the region, and for the well-being of the fishing communities themselves. Improving supply chain management is critical to building fishery resilience in the context of the COVID-19 pandemic. By investing in traceability, collaboration and communication, infrastructure, technology, and supporting small-scale fishers, the fishery sector can be made more resilient, reducing the risk of supply chain disruptions and ensuring that products are delivered to market in a timely and efficient manner.

The growth of the fishery sector over the next 30 years is expected to be influenced by a range of factors, including technological advancements, changes in consumer preferences, and the impact of pandemic diseases such as COVID-19. While the impact of pandemic diseases is likely to present significant challenges, these can be mitigated through effective disease control and resilience planning, helping to ensure the continued growth and stability of the fishery sector.

To ensure future resilience in the fishery sector, several key recommendations can be made

Invest in Sustainable Fishing Practices: Adopting sustainable fishing practices, such as using gear that minimizes bycatch and reduces the impact on the marine environment, can help to ensure the long-term viability of fish stocks and maintain the resilience of the fishery sector.

Promote aquaculture development: The expansion of aquaculture production can help to mitigate the impact of overfishing on wild fish stocks and reduce the risk of supply chain disruptions. By investing in research and development to improve the sustainability of aquaculture systems, the growth and resilience of the fishery sector can be increased.

Enhance food safety and traceability: Improving food safety and traceability can help to reduce the risk of food-borne illness and increase consumer confidence in seafood products. This can be achieved through the implementation of effective food safety management systems, the use of traceability technologies, and the establishment of robust food safety regulations.

Support small-scale fishers: Small-scale fishers play a critical role in many coastal communities, but often face significant challenges in accessing markets and getting their products to market in a timely and efficient manner. Supporting these fishers through initiatives such as training, technical assistance, and access to financing can help to build their resilience and ensure the continued growth of the fishery sector.

Strengthen supply chain management: Improving the efficiency and transparency of supply chain management is critical to reducing the risk of supply chain disruptions and ensuring that products are delivered to market in a timely and efficient manner. This can be achieved through the use of technologies such as blockchain and digital traceability systems, the development of robust infrastructure, and the promotion of collaboration and communication between actors in the supply chain.

Plan for pandemic resilience: The COVID-19 pandemic has highlighted the importance of having a resilient and effective food supply chain. To ensure future resilience in the fishery sector, it is important to invest in measures such as disease surveillance and control, food safety and traceability systems, and resilience planning. By adopting sustainable fishing practices, promoting aquaculture development, enhancing food safety and traceability, supporting small-scale fishers, strengthening supply chain management, and planning for pandemic resilience, the fishery sector can be made more resilient and better positioned to meet the challenges of the future.

References

Alam GMM, Sarker MdNI, Gatto M, Bhandari H, Naziri D (2022) Impacts of COVID-19 on the fisheries and aquaculture sector in developing countries and ways forward. Sustainability 14(3):1071

Azra MN, Kasan NA, Othman R, Noor GAGR, Mazelan S, Jamari ZB, Sar G, Ikhwanuddin M (2021) Impact of COVID-19 on aquaculture sector in Malaysia: findings from the first national survey. Aquac Rep 19:100568

Belton B, Rosen L, Middleton L, Ghazali S, Mamun AA, Shieh J, Noronha HS, Dhar G, Ilyas M, Price C (2021) COVID-19 impacts and adaptations in Asia and Africa's aquatic food value chains. Mar Policy 129:104523

Bennett NJ, Finkbeiner EM, Ban NC, Belhabib D, Jupiter SD, Kittinger JN, Mangubhai S, Scholtens J, Gill D, Christie P (2020) The COVID-19 pandemic, small-scale fisheries and coastal fishing communities. Coast Manag 48:336–347

Campbell SJ, Jakub R, Valdivia A, Setiawan H, Setiawan A, Cox C, Kiyo A, Darman, Djafar LF, Rosa E et al (2021) Immediate impact of COVID-19 across tropical small-scale fishing communities. Ocean Coast Manag 200:105485

Cashion T, Le Manach F, Zeller D, Pauly D, Frölicher TL (2017) Most fish destined for fishmeal production are food-grade fish. Fish Fish 18(5):837–844

Chuan OM, Ghazali A, Md Amin R, Bhubalan K, Nie LJ, Tuan Omar TMF, Khalil I, Assaw S, Chuen YJ, Mohd Yusoff N et al (2021) Positive and negative effects of COVID-19 pandemic on aquatic environment: a review. Sains Malaysiana 50:1187–1198

Clavelle T (2020, May 12) Global Fisheries During COVID-19—Global Fishing Watch. Global Fishing Watch. https://globalfishingwatch.org/data/global-fisheries-during-covid-19/

Coll M, Ortega-Cerdà M, Mascarell-Rocher Y (2021) Ecological and economic effects of COVID-19 in marine fisheries from the Northwestern Mediterranean Sea. Biol Conserv 255:108997

Cooke SJ, Twardek WM, Lynch AJ, Cowx IG, Olden JD, Funge-Smith S, Lorenzen K, Arlinghaus R, Chen Y, Weyl OLF et al (2021) A global perspective on the influence of the COVID-19 pandemic on freshwater fish biodiversity. Biol Conserv 253:108932

(2021) COVID-19 pandemic: challenges and recommendations. Chem Eng J 405:126683

Daw T, Gray T, Mombeshora B (2006) Stakeholder engagement in fisheries management: lessons learnt from the Lake Victoria fisheries management initiative. Afr J Ecol 44(3):303–312

De Silva RP, Karunarathne MDSD (2021) Fishery supply chain management: a review of recent trends, challenges, and prospects for sustainability. Ocean Coast Manag 207:105582

De Silva SS (2020) Fisheries in the future: the challenges ahead. Rev Aquac 12(1):1–22. https://doi.org/10.1111/raq.12328

EPA (2020) Frequent questions about disinfectants and coronavirus (COVID-19). Available online: https://www.epa.gov/coronavirus/frequent-questions-about-disinfectants-and-coronavirus-covid-19. Accessed on 29 July 2021

Fahlevi H, Chan S, Hasibuan P, Fadli N, Sofyan SE, Rianjuanda, Syukri M, Saidi T, Dawood R (2021) The feasibility of a cold storage facility for fish in Aceh during the COVID-19 pandemic. IOP Conf Ser Earth Environ Sci 674:012047

Fiorella KJ, Bageant ER, Mojica L, Obuya JA, Ochieng J, Olela P, Otuo PW, Onyango HO, Aura CM, Okronipa H (2021) Small-scale fishing households facing COVID-19: the case of Lake Victoria, Kenya. Fish Res 237:105856

Food and Agriculture Organization of the United Nations (FAO–COFI) (2021) Declaration for sustainable fisheries and aquaculture. Rome. https://doi.org/10.4060/cb3767en

Food and Agriculture Organization of the United Nations (FAO) (2016) The state of world fisheries and aquaculture. Contributing to food security and nutrition for all. Rome, p 200. License: CC BY-NC-SA 3.0 IGO

Food and Agriculture Organization of the United Nations (FAO) (2018) The state of the world fisheries and aquaculture—meeting the sustainable development goals. Rome, p 210. License: CC BY-NC-SA 3.0 IGO

Food and Agriculture Organization of the United Nations (FAO) (2022) The state of world fisheries and aquaculture 2022. FAO, Rome, p 226. https://population.un.org/wpp: Towards Blue Transformation. https://doi.org/10.4060/cc0461en

Food and Agriculture Organization of the United Nations (FAO) (2020a) Fishery and aquaculture statistics. Global production. Retrieve from http://www.fao.org/fishery/statistics/global-production/en

Food and Agriculture Organization of the United Nations FAO (2020b) The state of world fisheries and aquaculture. Sustainability in Action. Rome, p 244. License: CC BY-NC-SA 3.0 IGO

Food and Agriculture Organization of the United Nations FAO (2020c) Blockchain application in seafood traceability. Retrieved from http://www.fao.org/3/ca8945en/CA8945EN.pdf

Food and Agriculture Organization of the United Nations FAO (2020d) The effect of COVID-19 on fisheries and aquaculture in Asia. Bangkok. https://doi.org/10.4060/ca9545en

FAO (2020e) FAO report on overcoming water challenges in agriculture. https://www.fao.org/3/cb1447en/cb1447en.pdf

Gephart JA, Pace ML, Harfoot MB (2021) The impacts of climate change on fishery production: a global analysis. ICES J Mar Sci 78(1):33–42

Ghafoor D, Khan Z, Khand A, Ualiyevab D, Zaman N (2021) Excessive use of disinfectants against COVID-19 posing a potential threat to living beings. Toxicology 2:159–168

Hamzah A, Nurdin HS (2021) Economic resilience of fishermen community during COVID-19 pandemic. IOP Conf Ser Earth Environ Sci 715:012062

Hossain MS, Alam MA (2021) Sustainable fishery supply chain management: a review of challenges and opportunities. J Clean Prod 303:127183

Impacts of COVID-19 on the fisheries and aquaculture sector in developing countries and ways forward. Sustainability 14(1071):13

Islam MM, Khan MI, Barman A (2021) Impact of novel coronavirus pandemic on aquaculture and fisheries in developing countries and sustainable recovery plans: case of Bangladesh. Mar Policy 131:104611

Jalal KCA, John A, Abdullah R, Musa N, Ashraf MA (2021) Impact of Covid-19 on aquaculture and fisheries in ASEAN countries: some aspects of challenges, mitigations for future strategies in Malaysia. Desalin Water Treat 241:331–339

Kaewnuratchadasorn P, Smithrithee M, Sato A, Wanchana W, Tongdee N, Sulit VT (2020) Capturing the impacts of COVID-19 on the fisheries value chain of Southeast Asia. Fish People 18:2–8. Available online: http://repository.seafdec.org/handle/20.500.12066/6557

Knapp G, Roheim CA (2007) Poverty, consumption, and fisheries: recognizing and managing the multiple drivers of marine resource degradation. In: Human dimensions of global environmental change. Springer, Berlin, Heidelberg, pp 177–192

Knight CJ, Burnham TL, Mansfield EJ, Crowder LB, Micheli F (2020) COVID-19 reveals vulnerability of small-scale fisheries to global market systems. Lancet Planet Health 4(6):e219

Kumaran M, Geetha R, Antony J, Vasagam KPK, Anand PR, Ravisankar T, Angel JRJ, De D, Muralidhar M, Patil PK et al (2021) Prospective impact of Corona virus disease (COVID-19) related lockdown on shrimp aquaculture sector in India—a sectoral assessment. Aquaculture 531:735922

Lee J, Kim Y, Lee S (2019) Blockchain-based traceability in the seafood supply chain: a proposed system for small-scale fisheries in developing countries. Mar Policy 109:103695. https://doi.org/10.1016/j.marpol.2019.103695

Mandal SC, Boidya P, Haque MIM, Hossain A, Shams Z, Mamun AA (2021) The impact of the COVID-19 pandemic on fish consumption and household food security in Dhaka city, Bangladesh. Glob Food Sec 29:100526

Manlosa AO, Hornidge AK, Schlüter A (2021) Aquaculture-capture fisheries nexus under COVID-19: impacts, diversity, and social-ecological resilience. Marit Stud 20:75–85

Mejjad N, Cherif EK, Roder A, Krawczyk DA, El Kharraz J, Moumen A, Laqbaqbi M, Fekri A (2021) Disposal behavior of used masks during the COVID-19 pandemic in the Moroccan community: potential environmental impact. Int J Environ Res Public Health 18:4382

Miao Z, Zeng Q, Yang S (2022) A review of fishery supply chain management: current status and future directions. J Clean Prod 333:127073. https://doi.org/10.1016/j.jclepro.2020.127073

Mordor Intelligence (2021) Global fishery market—growth, trends, COVID-19 impact, and forecasts (2021–2026). Retrieve from https://www.mordorintelligence.com/industry-reports/global-fishery-market-industry

Ortega M, Mascarell Y, Coll M (2020) Monitoring of the COVID crisis in the Spanish Mediterranean fisheries. Weathering the storm of COVID 19 crisis. Fundacio ENT, p 10. https://ent.cat/wp-content/uploads/2020/12/2020_Seguimiento-crisis-COVID-y-sector-pesquero_EN.pdf

Ota Y, Bundy A, Roheim CA, Sumaila UR (2020) Towards resilient and sustainable global seafood supply chains. Nature Food 1(4):198–204

Paradis Y, Bernatchez S, Lapointe D, Cooke SJ (2021) Can you fish in a pandemic? An overview of recreational fishing management policies in North America during the COVID-19 crisis. Fisheries 46:81–85

Pilling GM, Guénette S (2019) Drivers of change in sustainable fisheries management. FAO Fisheries and Aquaculture Technical Paper (634)

Plagányi É, Deng RA, Tonks M, Murphy N, Pascoe S, Edgar S, Salee K, Hutton T, Blamey L, Dutra L (2021) Indirect impacts of COVID-19 on a tropical lobster fishery's harvest strategy and supply chain. Front Mar Sci 8:1–14

Ruiz-Salmón I, Fernández-Ríos A, Campos C, Laso J, Margallo M, Aldaco R (2021) Fishing and seafood sector in the time of COVID-19: considerations for local and global opportunities and responses. Curr Opin Environ Sci Health 23:100286

Salas S, Chuenpagdee R, Charles A (2020) Challenges in achieving sustainable small-scale fisheries in the global south. Curr Opin Environ Sustain 45:46–52

Sarkodie SA, Owusu PA (2021) Impact of COVID-19 pandemic on waste management. Environ Dev Sustain 23:7951–7960

Silva ALP, Prata JC, Walker TR, Duarte AC, Ouyang W, Barcelò D, Rocha-Santos T (2021) Increased plastic pollution due to COVID-19 pandemic: challenges and recommendations. Chem Eng J 2021(405):126683

Smith SL, Golden AS, Ramenzoni V, Zemeckis DR, Jensen OP (2020) Adaptation and resilience of commercial fishers in the Northeast United States during the early stages of the COVID-19 pandemic. PLoS ONE 15:e0243886

Song AM, Chuenpagdee R (2014) The economic viability of small-scale fisheries: a review. Mar Policy 45:41–48

Sunny AR, Mithun MH, Prodhan SH, Ashrafuzzaman M, Rahman SMA, Billah MM, Hussain M, Ahmed KJ, Sazzad SA, Alam MdT, Rashid A, Hossain MM (2021) Fisheries in the context of attaining Sustainable Development Goals (SDGs) in Bangladesh: COVID-19 impacts and future prospects. Sustainability 13(17):9912

Sveinsson H (2021) Aquaculture 2040—the blue revolution 2.0. A foresight study on sustainable growth in Norwegian aquaculture. Norwegian University of Life Sciences, Faculty of Environmental Sciences and Natural Resource Management

Waiho K, Fazhan H, Ishak SD, Kasan NA, Liew HJ, Norainy MH (2020) Potential impacts of COVID-19 on the aquaculture sector of Malaysia and its coping strategies. Aquac Rep 18:100450

White ER, Froehlich HE, Gephart JA, Cottrell RS, Branch TA, Agrawal Bejarano R, Baum JK (2021) Early effects of COVID-19 on US fisheries and seafood consumption. Fish Fish 22:232–239

Williams MJ, Kittinger JN (2017) A framework for identifying the capacity of small-scale fishing communities to adapt to the impacts of climate change. Sustainability 9(10):1834

Wiradana PA, Widhiantara IG, Pradisty NA, Mukti AT (2021) The impact of COVID-19 on Indonesian fisheries conditions: opinion of current status and recommendations. In: The 3rd international conference on fisheries and marine sciences. IOP Conf Ser Earth Environ Sci 718:012020. IOP Publishing

Workie E, Mackolil J, Nyika J, Ramadas S (2021) Deciphering the impact of COVID-19 pandemic on food security, agriculture, and livelihoods: a review of the evidence from developing countries. Curr Res Environ Sustain 2:100014

World Health Organization (WHO) (2020) Q&A on Coronaviruses (COVID-19). WHO, Geneva, Switzerland. Available online: https://www.who.int/emergencies/diseases/novel-coronavirus-2019/question-and-answers-hub/q-a-detail/q-a-coronaviruses. Accessed on 29 July 2021

Yilmaz I, Erdogan M (2021) Supply chain management practices in the Turkish fisheries industry. Int J Fish Aquat Stud 9(5):490–497

Yusoff FM, Abdullah AF, Aris AZ, Umi WAD (2021) Impacts of COVID-19 on the aquatic environment and implications on aquatic food production. Sustainability 13(11281):27

Zorriehzahra MJ, Hassantabar F, Ziarati M (2020) The impacts of COVID-19 pandemic on aquatic food production: a review. Iran J Aquat Anim Heal 6:15–22

Chapter 10
Life Cycle Assessment as Potential Tool to Gauge the Environmental Impact During Covid-19 Pandemic: Towards Better Waste Management

Noor Faizul Hadry Nordin, Husna Ahmad Tajuddin, Ahmed Jalal Khan Chowdhury, Osumanu Haruna Ahmed, Akbar John, and Bryan Raveen Nelson

Abstract COVID-19 pandemic hit the world since December 2019 and by 31st December 2022 it has achieved 651,918,402 cumulative cases with 6,656,601 cumulative deaths around the globe. During the COVID-19 pandemic, the application of Life Cycle Analysis (LCA) in waste management has been crucial to minimize long-term risks to human and environmental health. Studies suggested that there has been an increase in domestic and biomedical wastes produced during the COVID-19

N. F. H. Nordin (✉)
International Institute for Halal Research and Training, International Islamic University Malaysia, Level 3, KICT Building, 53100 Jalan Gombak, Selangor, Malaysia
e-mail: faizul@iium.edu.my

H. Ahmad Tajuddin
Department of Chemical Engineering and Sustainability, Kulliyyah of Engineering, International Islamic University Malaysia, 53100 Jalan Gombak, Selangor, Malaysia
e-mail: dr_husna@iium.edu.my

A. J. Khan Chowdhury (✉) · O. H. Ahmed
Faculty of Agriculture, Universiti Islam Sultan Sharif Ali (UNISSA), KM 33, Jalan Tutong, Kampung Sinaut, Tutong TB1741, Brunei Darussalam
e-mail: ahmed.chowdhury@unissa.edu.bn

O. H. Ahmed
e-mail: ahmed.haruna@unissa.edu.bn

A. John
College of Marine Science and Aquatic Biology, University of Khorfakkan, Sharjah, United Arab Emirates
e-mail: akbarjohn50@gmail.com

B. R. Nelson
Institute of Tropical Biodiversity and Sustainable Development, Universiti Malaysia Terengganu, 21300 Kuala Nerus, Terengganu, Malaysia
e-mail: bryan.nelson@umt.edu.my

D. Jones and A. J. Khan Chowdhury (eds.), *Environmental Resilience in the Pandemic Years 2020–2021*, Environmental Science and Engineering,
https://doi.org/10.1007/978-3-031-47757-7_10

pandemic, putting additional pressure on waste management systems, especially to developing countries with lack of effective infrastructure. The COVID-19 pandemic has revealed vulnerabilities in waste management chains, which could hinder disease containment and increase environmental pollution. The pandemic has led to an abrupt collapse of waste management chains, thus the application of LCA in waste management is crucial to prevent long-term risks. Several studies recommended fast adaptation through strengthening policies related to infectious waste, such as protection for waste collection workers and the application of LCA in waste management to prevent long-term risks. Additionally, the World Health Organization has called for reforms surrounding the disposal of medical waste, recommending eco-friendly packaging and reusable PPE to reduce plastic waste. This chapter is exploring for the potential of LCA approach in determining the environmental impact of different elements in waste management practices during the COVID-19 pandemic.

Keywords COVID-19 pandemic · Life cycle analysis · Waste management

10.1 Introduction

March 11th, 2020 marked as new historical event when global COVID-19 pandemic was announced by the WHO caused by SARS-CoV-2 virus. By 31st December 2022, Covid-19 have achieved 651,918,402 cumulative cases with 6,656,601 cumulative deaths around the globe (World Health Organization, n.d.) since it was firstly noticed at the end of December 2019 in Wuhan, PRC which said to be linked to an animal market that sell poultry, fish, and other animals to the public (Xu et al. 2020). Those infected with SARS-CoV-2 virus exhibit symptoms such as fever and cough, which were key sources of infection that transmitted via droplets and small airborne particles contaminated with the virus. It was also found out that the virus could be spread by asymptomatic carriers (Yu and Yang 2020). The asymptomatic carrier led to difficulty in controlling the spread of the virus since many not aware that they are the carrier. This led to many casualties, and many countries experienced an unexpected surge in the number of corpses that need to be managed by authorities with well-trained personnels equipped with proper PPE.

The economic breakdown due to the global impact of this virus towards various sector is inevitable (Nghiem et al. 2020). As a counter measure, every country implemented several precautionary actions such as imposing the physical distancing, imposing lockdown and movement control orders, closing their border, mandatory face masks usage and usage of gloves, aprons, and face shields in certain sectors in order to halt the spread of the virus. These measures were imposed in order to avoid their health system from collapse which underwent tremendous pressure from the alarming rate of increment in infected patients. As pandemic worsen, many movement control orders and lockdowns have been established in several nations as the prevention and control actions to stop the spreading of the virus. Around the world, a considerable increase in both municipal and medical waste has been documented during lockdowns. However, during this phase, even though working activities in offices and commercial areas were reduced, online food delivery and shopping were tremendously increased especially from residential areas. As a result, the contribution of the residential sector in generating the organic and inorganic waste has also increased which indirectly accountable for soil erosion and contamination towards water and air (Zambrano-Monserrate et al. 2020).

10.2 Impact of COVID-19 Pandemic to Waste Management

Solid waste is generated on daily basis in urban system which required a proper management for daily administration. Hence, solid waste management can be considered as essential services implemented by the local authorities to ensure the cleanliness of the residential areas thus maintain the hygienic of the surrounding areas (Hoornweg and Bhada-Tata 2012). In the presence of natural disaster and pandemics, the role and function of solid waste management becomes much more critical. During the pandemic, the drastic change in waste generated was evidence. Face masks, other PPEs and hand sanitizers are now routinely used in our daily lives. The usage of these items leads to addition of waste generated thus increase significant volume load to waste treatment systems. In order to avoid disease transmission, most of the countries implemented the lockdown, movement control order and work from home policies. As a result, the increase of solid waste generated from residential sectors was evidenced due to these actions. In addition, those who had mild COVID-19 symptoms or asymptomatic are recommended for home surveillance and quarantine due to significantly high increase of positive cases that overwhelmed the existing medical facilities. The home quarantine produced medical waste such as tissues, body fluids and PPEs that required careful disposal to avoid the potential risk of viral transmission to other family members and waste collectors (Mol and Caldas 2020). Every individual is also required to wear a face mask to avoid the spreading of the virus through droplets and aerosols, thus resulted in the increment of the waste load (Bauchner et al. 2020). Furthermore, the recommendation from health authorities for not reusing the face mask due to the substantial risk of viral retention also resulted in significantly increase in demand for PPE thus increase the issue up to the crisis mode. There are also concern on improper disposal of facemasks and PPE which may landed into the streams and rivers has led in an explosion of plastic waste contamination since most of the face mask were produced with hydrophobic protective layer consist of polypropylene (PP) and also polyurethane (PUR) or polyacrylonitrile (PAN) in other expensive type of face masks (Konyn 2020) to avoid the droplets of body fluid. Furthermore, the virus on the facemasks may enter the aquatic environment, potentially infecting those exposed to it.

Table 10.1 summarizes the global usage of face mask during the COVID-19 pandemic. Research shows that pandemic has created a paradigm shift in waste generation forms such as unexpected jump in the volume load of plastic wastes from the usage of food packaging and single-use plastic as people are trying to avoid recycles materials for their safety. There is high probability that suspicious over the safety of recyclables are one of the factors that lead to the ascending use of single-use plastics. The usage of personal or reusable containers was also not allowed in most restaurants and cafe but instead, the single-use packaging materials were preferred (Naughton 2020).

On the other hand, the increase of food consumption at home and online shopping were also increased due to movement restriction order and work from home policy, thus contributed tremendously to household waste. As instance, in just 56 days of implementation of lockdown in Singapore, an extraneous of 1400 tonnes of plastics was produced (Bengali 2020). With the volume of plastic garbage increasing dramatically, this restriction has placed further burden on waste management systems. Additionally, according to a survey conducted by Shammi et al. (2021), more than 49% of people disposed their home waste and COVID-19 related waste in the same container, which may exacerbate infection. The developing countries faced greater risk of contamination from the waste and wastewater compared to developed nations. In most developing countries the waste was disposed in improper landfills and dumping sites where regularly visited by ragpickers without proper safety equipment. Other than ragpickers, the janitors and healthcare personnels are also vulnerable to contaminated household and medical waste thus possessed high risk to be infected and ultimately hamper the area's cleanliness. As findings suggested that the virus may potentially be infectious for several days on surfaces, improper handling of wastes can be a significant contribution in the spreading of the disease.

Kampf et al. (2020) demonstrated that SARS-CoV-2 virus can survive and remain active on hard surface such as plastic and metal up to 9 days, hence raising concern in many nations that it will increase the possibility of contamination to general waste especially when the efficient solid waste management does not exist and lead to numerous issues on environmental and health (Yu et al. 2020). The amount of COVID-19 related medical waste also increased significantly with the accelerated number of positive cases. Due to the persistency of SARS-CoV-2 in the environment, the confirmed patient contributed to tremendous environmental contamination and the control of its source of infection by effective management of medical waste is the only way to solve it. The severity of medical waste generation as a result of

Table 10.1 The estimated usage of mask used during the COVID-19 pandemic, estimation was demonstrated by Nzediegwu and Chang (2020)

Continents	Total daily facemask used (million)	Weight of total daily facemask used (tonnes)
Asia	3716.2	1486.48
Africa	922.22	368.89
Europe	884.71	353.88
North America	489.05	195.62
Oceania	45.43	18.17
South America	544.39	217.75

this pandemic have been shown by several studies such as by Peng et al. (2020) who demonstrated that China generated approximately 469 tonnes of medical waste every day. Mihai (2020) also mentioned that Indonesia generated about 12,740 tonnes of medical waste within the first 3 months of COVID-19 discovery with the usage of PPE as the major contribution. This became global attention since the used PPE is the source of plastic pollution and subsequently will become major contributor for ongoing microplastic pollution. Making it worst, when many developing and emerging countries were not prepared and the lack of appropriate guidelines for disposing of these wastes will lead to the potential for the spread of virus transmission in the future (Klemeš et al. 2020). Improper medical waste management will only magnify the disease transmission problem with during and even after the COCID-19 pandemic period. With the setup of the isolations and quarantine centres as addition of all hospitals and medical centres, increased the burden to medical waste management system with vast amount of medical waste. This situation is globally alarming especially to developing and undeveloped countries, as numerous hospitals do not possess sufficient and fail to implement proper disposal techniques (Shammi and Tareq 2021).

Despite incineration being the most effective method of disposing medical waste, a number of hospitals in developing nations lack the necessary facilities for it. As indicated by Shammi and Tareq (2021), many hospitals in Bangladesh burn their medical wastes in the backyard due to inaccessibility to proper facilities thus further raises the occurrence of air pollution. The press conference for China's State Council's joint prevention and control mechanism revealed that approximately 468.9 tons of medical waste related to COVID-19 is produced daily (SCIO PRC 2020). Hence, to effectively combat the pandemic, it is crucial to dispose of medical waste related to COVID-19 in a prompt, organized, efficient, and safe manner. In order to achieve that, some guidelines to handle medical related waste during covid-19 pandemic had been established:

a. **Classified packaging and pre-treatment**
    i. The hospital provided designated bins in public areas for the disposal of medical waste, particularly for the discarded face masks. The double-layered medical waste bags containing the wastes should be disposed of by a designated personnel as general medical waste.
    ii. Waste generated from medical and general sources from any medical facilities related to COVID-19 and as well as other COVID-19 related laboratories especially for SARS-CoV-2 detection analysis, ought to be considered as medical waste linked to COVID-19. The waste should also be labelled as "COVID-19 infection".
    iii. Double layered medical waste bag should be used for packaging of infectious and pathological waste. Subsequently, chlorine containing disinfectant should be used for sterilization by applying a spray to the bags before placing them in a closed container for medical waste.
    iv. The sharps wastes should be placed into dedicated container to prevent any impairment to the packing. Then, the box will be double-layered packed and

should be disinfected with a solution containing chlorine before disposing of the medical waste. Before disposing of packed or sealed specimens and preservation solutions containing pathogens in the medical waste bin, they must be autoclaved for 110 min at 121 °C.

b. **Temporary Storage**

   i. A designated area for storing COVID-19 medical waste should be created, which is clearly marked with warning signs. Medical waste container from medical facilities dealing with COVID-19 and testing laboratories should be provided in this place to prevent them from being mixed with regular hospital waste.
   ii. Medical waste related to COVID-19 should not be kept in this temporary storage area for more than 24 h.

c. **Collection and transportation**

   i. Trained personnel and separate vehicles should handle the disposal of medical waste related to COVID-19, and these should not be mixed with general medical waste.

d. **Disinfect the storage area**

   i. The transportation path ought to steer clear of congested area as much as is feasible, and the timing should avoid the busy periods during the morning and evening commutes.
   ii. As soon as loading and unloading are completed, it is important to disinfect the storage area and vehicles.

e. **Centralized disposal**

   - Due to its ability to effectively eliminate infectious pathogens and its suitability for different types of medical waste, high-temperature incineration is the preferred method of treating medical waste associated with COVID-19.
   - However, hospitals and institutions in less developed countries or medium- and low-risk areas were not able to apply it.
   - Sanitary landfill is also a viable option for disposing of medical waste after been autoclaved if incinerator was not available.

10.2.1 The Increased of Waste Volume During COVID-19 Pandemic

Developing countries with higher population having a crisis during COVID-19 not only for regulating the spread of the COVID-19, but also to manage the larger amount of waste made up of materials like solid, plastic, and medical products. In correlation with the rise in the quantity of individuals who are infected, isolated, or put into quarantine, the massive amount of biomedical waste generated from all these patients also

Table 10.2 The expected production of medical waste in ASEAN nations based on the pattern observed in China (Tripathi et al. 2020)

City, Country	Population (World population review) (number)	Additional medical waste (kg/day)	Total waste production over 60 days (kg)
Manilla, Philippines	14×10^6	280×10^3	16.8×10^9
Jakarta, Indonesia	10.6×10^6	212×10^3	12.75×10^9
Kuala Lumpur, Malaysia	7.7×10^6	154×10^3	9.24×10^9
Bangkok, Thailand	10.5×10^6	210×10^3	12.6×10^9
Hanoi, Vietnam	8×10^6	160×10^3	9.6×10^9

*The amount of medical waste generated is expected to be 3.4 times from the number of people who have been infected

increased significantly. It was estimated that infected patients contributed about 81%, while approximately 16% of the waste produced was due to the patients who were under quarantine. In general, the volume of masks and gloves generated is in correlation with the increasing number of the infected patients. Other than number of infected patients, factors such as urban population percentage and total number of populations may also affect the total utilization of masks and gloves for protection. The global vaccination program also give impact to waste management system. The usage of vaccination kits globally was also raised the concerns among environmentalist due to their contribution to waste management system.

According to Celis et al. (2021), for each real-time PCR test produces about 37.27 g of plastic waste. Hence, the amount of hospital and laboratory waste has significantly risen due to increase of COVID-19 testing. Table 10.2 indicates the expected increment of hospital and laboratory wastes in ASEAN countries amidst the outbreak of COVID-19 (Tripathi et al. 2020).

10.2.2 Waste Management Plan Amidst the Outbreak of COVID-19

Understanding the lifecycle of the SAR-CoV-2 virus, the transmission and its control pathways are the key factors to properly and safely manage the waste generated from residential areas, streets and health facilities during the pandemic. The main pathway of SARS-CoV-2 virus transmission in the waste management system is from the waste generators to the waste collectors as the virus can remain viable on a solid surface for several days (Nghiem et al. 2020; Scheinberg et al. 2020). The possibility of virus release through aerosol is al so high when compaction system is used in waste collection systems for waste collected from the house of quarantined patients (Mol and Caldas 2020). The virus can survive for longer duration on the smooth surfaces, and it can be varied on other different materials. The persistence of the SARS-CoV-2 virus found on inert materials can be up to 9 days and the time length is reduced with temperature higher than 30 °C. In addition, 70% ethanol, 0.5% hydrogen peroxide and 0.1% bleach solution can kill the virus within 1 min (Kampf et al. 2020). Other common disinfectants such as iodophors and phenolics can also be used to kill the SARS-CoV-2 virus (Wang et al. 2020). The 80% ethanol or 75% 2-propanol were also demonstrated as two disinfectants recommended by WHO for their suitability during viral outbreaks (Siddharta et al. 2017). The 5 min exposure to 7.5% Povidone-iodine, 0.1% Benzalkonium chloride, 0.05% Chloroxylenol, and Chlorhexidine were demonstrated able to deactivate the virus while it took about 15 min for 1:49 hand soap solution to be effective (Chin et al. 2020).

Several protocols and guidelines were outlined for the management of waste during pandemics. For instance, The Association of Cities and Regions for Sustainable Resource Management (ACR+), Brussel provides protocols for safety of frontline waste workers in collecting waste from COVID-19 patient's house and quarantine facilities. Similarly, the Public Health of England (PHE) provides similar guideline for self-quarantine patients where the used disposable items should be disposed in the rubbish bag double layered with another disposable bag. Like many other European countries, Ministry of Environment of the Czech Republic requires items from quarantine centre or household of SARS-CoV-2 infected person must be sealed properly using a plastic bag that is 0.2 mm thick, and if a thinner bag is used, then it should be double packed. From Central Pollution Board (CPCB) India, the guideline stated that biomedical waste from isolation ward should be labelled as COVID-19 waste and separately handled. It is recommended that the containers be sanitized daily using a solution of sodium hypochlorite at a concentration of 1%. It is crucial to disinfect the waste before collecting it to minimize the chances of infection for the collection workers.

Solid waste management was already a difficult task prior to the outbreak of COVID-19 for many developing countries where it is not fully developed and dynamic. The addition of COVID-19 waste, creates an extra burden to be handled to the existing solid waste. Even though the most effective treatment for these wastes is through incineration, many developing countries does not have suitable incineration facilities. For instance, countries like Bangladesh having problem in battling with the COVID-19 wastes due to the lack of suitable infrastructures thus lead to high percentage of mismanaged waste (Shammi and Tareq 2021). The most suitable approach to effectively deal with the large quantity of waste in this scenario is to enforce a just and productive COVID waste management strategy. According to the world bank classification, only 20% of total waste being recycled in the low-income countries while most of the generated wet waste will be dumped openly. In comparison, due to better waste management system and facilities, only 51% of the waste in high-income countries end up to the secured landfill since the majority of the waste was recycled.

As part of waste management system, waste recycling also been affected by the COVID-19 pandemic. During COVID-19 Pandemic, some countries like USA, the waste recycling rate was reduced in some of the cities because of high potential risk from the spreading of the SARS-CoV-2 virus. During the lockdown, several cities and states halted their recycling program (Liu et al. 2020). This reduction can be seen in California, Michigan and Oregon, where their recycling facilities facing challenges due to plummeted collection of recyclable materials (Staub 2020). The disposal of these enormous amounts of wastes also posed a severe challenge due to the decline number of workers by half because of fear to SARS-CoV2 transmission. Decline in attendance of employees at work affects the proper functioning of recycling facilities thus impact the nations through the lack of returns. According to Plastic Recycler Europe, since companies involved with waste were encouraged to incinerate the recyclable waste amidst the COVID-19 outbreak, the many European companies in recycling industry also experiencing a major crisis due to the reduction of resource recovery (recyclable plastic) and subsequently lead to adverse environmental impacts. As example, Vienna experienced approximately 10–15% reduction in material collected for recycling.

Some countries introduced new initiatives to maintain the safety of waste collection workers and enhance the waste recycling industries. For instance, two bonus schemes were introduced by Hong Kong government to assist the recycling industries (HKPC 2020):

i. One-off Rental Support Scheme (ORSS): Recycling centres are given 50% discount for their rental payment (or up to HKD $25,000).
ii. One-off Recycling Industry Anti-Epidemic Scheme (ORIAS): HKD $20,000 monthly are given to the recycling centres to assist their cost of operation.

To ensure the safety of the employees, Vienna also implemented some strict measures. The number of streets sweeping workers were reduced to 50% and diverted to more needed services like waste collection. Office staff and personnels were kept to minimum numbers and encourage to work from home. The workers present at the location were given instruction to have proper Personal Protective Equipment (PPE) and were obligated to adhere to the standard procedure where they need to frequently washing their hands and maintaining a one-meter distance from others while performing their duties. These similar standard procedures were also implemented by other countries in European Union and Canada. International Solid Waste Association (ISWA) has established three objectives for the management of waste in the midst of the COVID-19 pandemic:

i. The proper handling of waste should be upheld globally without exception. Additionally, it is important to provide waste management personnel with adequate protective equipment to safeguard their health and well-being.
ii. To prevent infection or cross-contamination, it is necessary for the recycling process to be restated.
iii. To ensure safety, the proper disposal of biomedical waste should be conducted, following the temporary guidelines produced for the municipal waste service providers, generator-citizen, COVID-19 patients and companies and operators involved in recycling.

As the COVID-19 pandemic bring the major changes in waste management system, further analysis related to the changes and its suitability to curb the SARS-CoV-2 virus is needed. One of the techniques to determine the effectiveness of these changes especially towards environmental impact is Life cycle assessment (LCA).

10.3 Life Cycle Assessment

A Life Cycle Assessment (LCA) is defined as the systematic analysis of the potential environmental impacts of products or services during their entire life cycle. During a Life Cycle Assessment (Life Cycle Analysis), the assessment of the probable effects on the environment at every stage of a product or service's life cycle, which includes production, distribution, utilization, and disposal, will be conducted. The aforementioned encompasses the production procedures such as the production of basic materials, auxiliary, and operating materials, as well as the use and disposal stages such as waste incineration, along with the upstream and downstream processes such as suppliers and waste management, respectively. In LCA, it is important to take into account the main types of environmental effects, which are related to the utilization of resources, human well-being, and the impact on the ecosystem. Since an LCA evaluates the effects of a product throughout its life cycle, beginning with the extraction of raw materials and ending with disposal, it is occasionally known as a "(Klopffer and Grahl 2014). The Life Cycle Assessment (LCA) stages diagram is illustrated in Fig. 10.1.

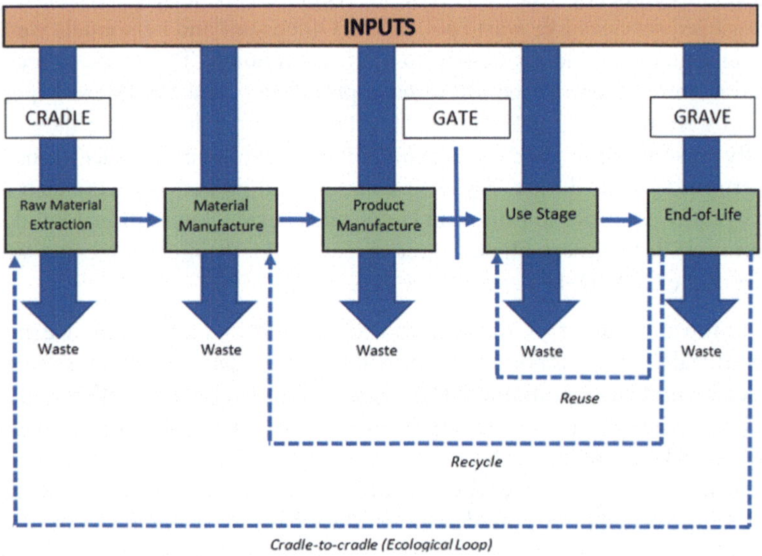

Fig. 10.1 The Life Cycle Assessment (LCA) system boundary

In principle, the Life Cycle Assessment (LCA) will use guidelines and requirements according to ISO 14040 which provides the principles and framework for managerial audience and ISO 14044 for the practitioners as provide by the International Organization for Standardization (ISO). LCA is a method used to evaluate the environmental elements and probable effects linked with a product, procedure, or facility by:

i. Inventory: Creating a register list of important energy and material inputs, as well as environmental emissions.
ii. Impact analysis: Evaluating the potential environmental implications of selected inputs and outputs.
iii. Interpretation: Interpreting the findings to assist in making a more informed decision.

There are two main types of LCA; (1) Attributional LCAs that seeks to attribute the load linked to the creation and consumption of a commodity, or relating to a particular process or method, for an identified temporal period (Gong and You 2017) and (2) Consequential LCAs that seek to determine the impacts on the environment that may result from a decision or a proposed modification within a system being examined, and thus focusing on the future and necessitating consideration of market and economic implications (Gong and You 2017). In other words, Attributional LCA "attempts to answer, 'how are things (i.e., pollutants, resources, and exchanges among processes) flowing within the chosen temporal window?', while Consequential LCA attempts to answer 'how will flows beyond the immediate system change in response to decisions?" (USEPA 2006).

Life Cycle Analysis (LCA) is a tool that can be implemented in managing the solid waste amidst the COVID-19 pandemic to assess the environmental impact of various waste management practices. This information can then be used to make informed decisions about which practices are most sustainable and efficient. LCA able to provide data-driven decision for solid waste management agencies to promote sustainability and reduce the negative effects that the COVID-19 pandemic has on the environment. Some of the ways where LCA can be utilized include:

i. Assessing the environmental impact of waste collection and transportation methods, such as the utilization of personal protective gear and the impact of reduced waste collection frequency.
ii. Evaluating the environmental impact of different disposal methods, such as landfilling, composting, and incineration and to determine the best method for managing waste amidst the COVID-19 outbreak.
iii. Identifying ways to reduce waste generation, such as promoting telecommuting and reducing single-use packaging and to reduce the amount of waste that needs to be managed.
iv. Evaluating the potential impact of changes in consumer behaviour, such as increased online shopping and reduced dining out, on the waste stream.

Life Cycle Analysis (LCA) can be used to compare different treatments for biomedical waste amidst the COVID-19 outbreak by evaluating the environmental impact of each treatment option. In general, there are four phases in an LCA; goal and scope definition, life cycle inventory analysis (LCI), life cycle impact assessment (LCIA) and life cycle interpretation (ISO 14040 2006) (Fig. 10.2). Life cycle analysis (LCA) methodology for solid waste management involves the steps as illustrated in Fig. 10.3.

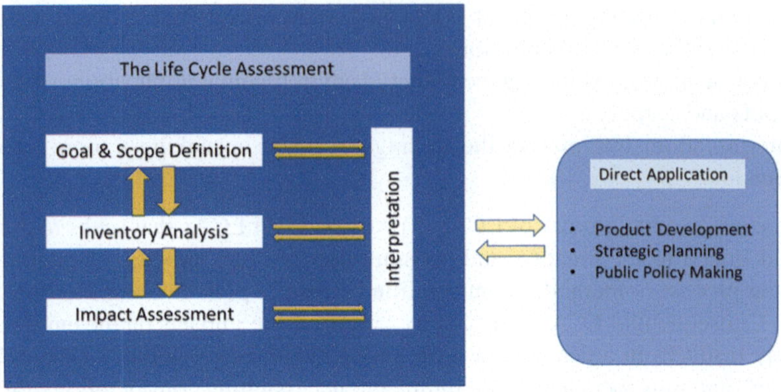

Fig. 10.2 Phases of a Life Cycle Assessment (ISO 14040 2006)

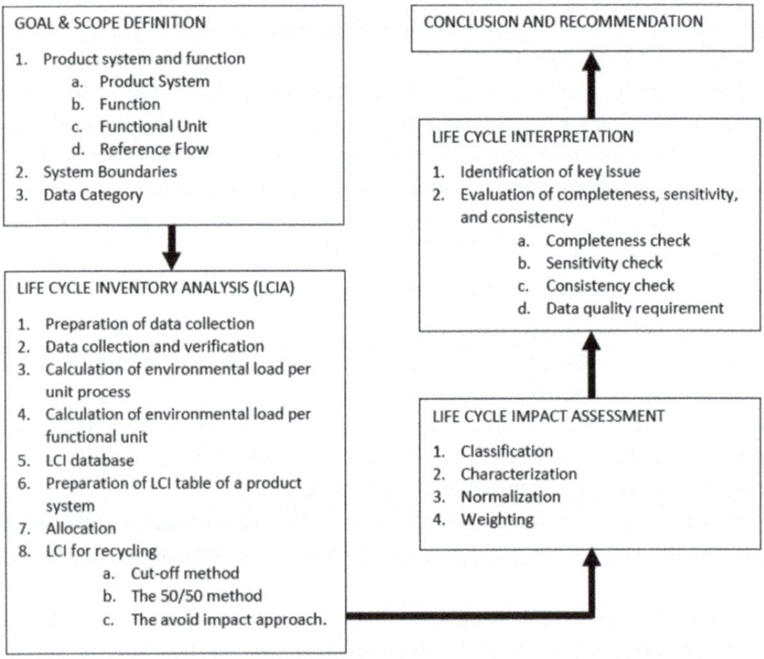

Fig. 10.3 Overall methodology for life cycle analysis in solid waste management

10.3.1 Application of LCA for Waste Management During COVID-19 Pandemic

COVID-19 pandemic exhibits unusual burden to the waste management facilities due to the rise in the production of both medical and household waste from health facilities and residential areas, respectively. The usage of protective gear kits which possibly having the ability to spread an infection need to be handled and disposed properly to avoid the risk posed to human well-being, as well as the ecological systems of both land and sea. Given the various technology available, Kumar et al. (2021) utilised Life Cycle Assessment using cradle to grave analysis for PPE kits disposal options in India. The research examined every stage of PPE kit production and use, from obtaining raw materials to disposal, using Life Cycle Assessment tools. Its aim was to turn the waste management industry into a secondary resource recovery sector and integrate it with manufacturing to promote a circular economy and conserve ecosystem services using a life cycle approach. Several disposal methods such as landfill, centralized incinerator and decentralized incinerator were analysed. From the Life Cycle Analysis conducted, it is indicated that decentralized incineration appears to be a feasible alternative for PPE disposal, taking into account both environmental and health considerations. Landfill-based disposal is the least feasible choice due to its negative effects on all impact categories, except for Global Warming Potential (GWP). Compared to centralized incinerators, the decentralized incinerator has a smaller impact on environmental indicators such as EP, HTP, POCP, and GWP. Simultaneously, it generates nearly identical effects with regards to AP and FAETP. Given the above results, it is important to create strategies of handling such type of wastes in advance given the times that changing fast, and policy decisions that to be taken with speed and scientific accuracy to reduce the impact on human lives. The LCA approach in the present work has demonstrated that it can be used as an important tool in such decision making and that environmentally sound and sustainable strategies can be devised using it.

Other than PPE kits, medical waste generated by COVID-19 patients from health facilities and quarantine centres were also exponentially increased during the pandemic. These infectious wastes require efficient and urgent disposal for preventing the pandemic spread. Facilities for emergency disposal of biomedical waste must have criteria such as convenient movement, easy installation, and operation with efficient energy supply that able to quickly eradicate or minimize the infection severity of biomedical waste. Zhao et al. (2021) conducted comparative analysis for three movable disposal scenarios and two co-incineration scenarios for emergency disposal during the COVID-19 pandemic in China. The environmental impacts and key factors for three scenarios of movable disposal were determined by Life Cycle Assessment. The three movable disposal scenarios are incineration disposal vehicle, movable steam followed by co-incineration with municipal solid waste and microwave sterilization followed by co-incineration with municipal solid waste. The results show that co-incineration with municipal solid waste has the lowest environmental impacts due to environmental benefits produced by power generation, while co-incineration with hazardous waste is the highest due to the high energy consumption. Energy consumption from kerosene, electricity and diesel are the key factors for three movable disposal scenarios. For movable steam and microwave sterilization followed by co-incineration with municipal solid waste, power generation from incinerating disinfected medical waste has significant beneficial environmental impacts since it able to avoid impacts of electricity consumption.

10.4 Conclusion

The COVID-19 pandemic has had a significant impact on waste management systems globally. With more people staying at home, there has been a significant increase in the amount of household waste generated, including packaging materials, food waste, and personal protective equipment (PPE) such as masks and gloves. The pandemic also has caused disruptions to waste collection and processing systems, as waste management workers have been deemed essential workers and have continued to work during lockdowns and other restrictions, but disruptions in supply chains, changes in consumer behaviour, and increased safety measures have made it more difficult and expensive to manage waste. With the increase in online shopping and food delivery services, there has been a shift towards single-use packaging, leading to a further increase in waste generation. The widespread use of PPE has resulted in a significant increase in the amount of medical waste that needs to be managed, and proper disposal of this waste is important to prevent the spread of COVID-19. Overall, the COVID-19 pandemic has highlighted the importance of having resilient and effective waste management systems in place, as well as the need to reduce waste and shift towards more sustainable forms of consumption and production.

Life Cycle Assessment (LCA) can play a role in improving waste management during the COVID-19 pandemic. By using LCA, waste management professionals and decision-makers can identify areas for improvement in the waste management system and make informed decisions about the best ways to manage waste during the pandemic. LCA can be used to evaluate the environmental impact of different waste management options, such as recycling, composting, or incineration. This information can help decision-makers prioritize the most environmentally friendly options, which can be especially important during the pandemic when there may be increased pressure to quickly dispose of waste. LCA can also be used to assess the impact of changes in consumer behaviour and the increased use of single-use packaging during the pandemic. This information can help decision-makers develop strategies to reduce the amount of waste generated, such as promoting reusable products, reducing packaging, and increasing recycling efforts. In conclusion, LCA can provide valuable insights into the environmental impact of waste management during the COVID-19 pandemic and help decision-makers make informed decisions about the best ways to manage waste.

References

Bauchner H, Fontanarosa PB, Livingston EH (2020) Conserving supply of personal protective equipment—a call for ideas. JAMA 323(19):1911–1911

Bengali S (2020) The COVID-19 pandemic is unleashing a tidal wave of plastic waste. The Los Angeles Times

Celis JE, Espejo W, Paredes-Osses E, Contreras SA, Chiang G, Bahamonde P (2021) Plastic residues produced with confirmatory testing for COVID-19: classification, quantification, fate, and impacts on human health. Sci Total Environ 760:144167

Chin AW, Chu JT, Perera MR, Hui KP, Yen HL, Chan MC, Peiris M, Poon LL (2020) Stability of SARS-CoV-2 in different environmental conditions. Lancet Microbe 1(1):e10

Gong J, You F (2017) Consequential life cycle optimization: general conceptual framework and application to algal renewable diesel production. ACS Sustain Chem Eng 5(7):5887–5911

Kumar H, Azad A, Gupta A, Sharma J, Bherwani H, Labhsetwar NK, Kumar R (2021) COVID-19 creating another problem? Sustainable solution for PPE disposal through LCA approach. Environ Dev Sustain 23:9418–9432. https://doi.org/10.1007/s10668-020-01033-0

Hong Kong Productivity Council (HKPC) (2020) Plastics recyclers cease production. https://www.plasticsrecyclers.eu/post/plastics-recyclers-cease-production

Hoornweg D, Bhada-Tata P (2012) What a waste: a global review of solid waste management. Urban development series; knowledge papers no. 15. World Bank, Washington, DC. © World Bank. https://openknowledge.worldbank.org/handle/10986/17388

ISO 14040:2006 Environmental management—Life cycle assessment—Principles and framework

Kampf G, Todt D, Pfaender S, Steinmann E (2020) Persistence of coronaviruses on inanimate surfaces and their inactivation with biocidal agents. J Hosp Infect 104(3):246–251

Klemeš JJ, Van Fan Y, Tan RR, Jiang P (2020) Minimising the present and future plastic waste, energy and environmental footprints related to COVID-19. Renew Sustain Energ Rev 127:109883

Klopffer W, Grahl B (2014) Life cycle assessment (LCA). Wiley-VCH Verlag GmbH & Co. KGaA, pp 1–2

Konyn C (2020) Another side effect of COVID-19: the surge in plastic pollution. Earth.Org—past, present, future [Online]. Available https://earth.org/covid-19-surge-in-plastic-pollution/

Life Cycle Assessment (LCA) (2012) EPA. gov. Washington, DC. EPA National Risk Management Research Laboratory (NRMRL). https://web.archive.org/web/20120306122239/ http://www.epa.gov/nrmrl/std/lca/lca.html

Liu M, Tan S, Zhang M, He G, Chen Z, Fu Z, Luan C (2020) Waste paper recycling decision system based on material flow analysis and life cycle assessment: a case study of waste paper recycling from China. J Environ Manage 255:109859

Mihai FC (2020) Assessment of COVID-19 waste flows during the emergency state in Romania and related public health and environmental concerns. Int J Environ Res Public Health 17(15):5439

Mol MPG, Caldas S (2020) Can the human coronavirus epidemic also spread through solid waste? Waste Manage Res 38(5):485–486

Naughton CC (2020) Will the COVID-19 pandemic change waste generation and composition? The need for more real-time waste management data and systems thinking. Resour Conserv Recycl 162:105050

Nghiem LD, Morgan B, Donner E, Short MD (2020) The COVID-19 pandemic: considerations for the waste and wastewater services sector. Case Stud Chem Environ Eng 1:100006

Nzediegwu C, Chang SX (2020) Improper solid waste management increases potential for COVID-19 spread in developing countries. Resour Conserv Recycl 161:104947

Peng J, Wu X, Wang R, Li C, Zhang Q, Wei D (2020) Medical waste management practice during the 2019–2020 novel coronavirus pandemic: experience in a general hospital. Am J Infect Control 48(8):918–921

Scheinberg A, Woolridge A, Humez N, Mavropoulos A, Silva C, Savino A, Ramola A (2020) Waste management during the COVID-19 pandemic. International Solid Waste Association, p 12

Shammi M, Behal A, Tareq SM (2021) The escalating biomedical waste management to control the environmental transmission of COVID-19 pandemic: a perspective from two south Asian countries. Environ Sci Technol 55(7):4087–4093

Shammi M, Tareq SM (2021) Environmental catastrophe of COVID-19: disposal and management of PPE in Bangladesh. Glob Soc Welfare 8:133–136

Siddharta A, Pfaender S, Vielle NJ, Dijkman R, Friesland M, Becker B et al (2017) Virucidal activity of World Health Organization—recommended formulations against enveloped viruses, including Zika, Ebola, and emerging coronaviruses. J Infect Dis 215(6):902–906

Staub C (2020) COVID-19 damages supply chain for deposit materials—resource Recycling News [Online]. Available https://resource-recycling.com/recycling/2020/04/14/covid-19-damages-supply-chain-for-deposit-materials/

The State Council Information Office of the People's Republic of China (SCIO) (2020) Press conference on strengthening comprehensive treatment of medical waste and environmental protection. Available at http://www.scio.gov.cn/xwfbh/gbwxwfbh/xwfbh/wsb/Document/1675174/1675174.html

The World Bank. Trends in solid waste management. https://datatopics.worldbank.org/what-a-waste/trends_in_solid_waste_management.html

Tripathi A, Tyagi VK, Vivekanand V, Bose P, Suthar S (2020) Challenges, opportunities and progress in solid waste management during COVID-19 pandemic. Case Stud Chem Environ Eng 2:100060

USEPA (2006) Life cycle assessment: principles and practice. U.S. Environmental Protection Agency, Cincinnati, Ohio, pp 3–9

Wang J, Shen J, Ye D, Yan X, Zhang Y, Yang W, Li X, Wang J, Zhang L, Pan L (2020) Disinfection technology of hospital wastes and wastewater: suggestions for disinfection strategy during coronavirus disease 2019 (COVID-19) pandemic in China. Environ Pollut 262:114665

World Health Organization (n.d.) Who coronavirus (COVID-19) dashboard. World Health Organization. Retrieved 9 Mar 2023, from https://covid19.who.int/

Xu X, Chen P, Wang J, Feng J, Zhou H, Li X, Zhong W, Hao P (2020) Evolution of the novel coronavirus from the ongoing Wuhan outbreak and modeling of its spike protein for risk of human transmission. Sci China Life Sci 63:457–460

Yu X, Yang R (2020) COVID-19 transmission through asymptomatic carriers is a challenge to containment. Influenza Other Respir Viruses 14(4):474–475

Yu H, Sun X, Solvang WD, Zhao X (2020) Reverse logistics network design for effective management of medical waste in epidemic outbreaks: insights from the coronavirus disease 2019 (COVID-19) outbreak in Wuhan (China). Int J Environ Res Public Health 17(5):1770

Zambrano-Monserrate MA, Ruano MA, Sanchez-Alcalde L (2020) Indirect effects of COVID-19 on the environment. Sci Total Environ 728:138813

Zhao H, Liu H, Wei G, Wang H, Zhu Y, Zhang R, Yang Y (2021) Comparative life cycle assessment of emergency disposal scenarios for medical waste during the COVID-19 pandemic in China. Waste Manage 126:388–399

Chapter 11
The Environmental Impacts and Transformation on the Tourism During COVID-19

Zhongxian Jiang, Huanna Zhao, Fa Yun, and Wenting Jiang

Abstract Different festival environments and transformations may provide conceptual resources for transforming tourism into more responsible and sustainable practices. Considering together the scientific methodology of two typically disparate 'environments', and Transformation Tourism, the COVID-19 pandemic was also treated as a spatial–temporal beginning for the transformation of the tourism industry. This study was also conducted to determine tourist behavior after the Covid-19 pandemic by identifying tourist typology and tourist travel motivations. A questionnaire-based survey method was used. An online survey was done, and responses were obtained from various tourist destinations of China. This approach also helps dismantle the mechanisms that perpetuate the pernicious aspects of host-host disparities in tourism. After the reopening of borders, starting of public mobility, and resumption of recreational activities, the transformative ability of festivals to transform their participants, both in and out of highly perforated times, is not to the extent that the distinction between guest and host is completely erased. The creative and community responses of such transformative festival cultures and past crises were linked to participation, civic responsibility, etc. Values were presented as an example of how people can help overcome shared emergencies and provision of further sustainable tourism during the COVID-19. Such a destructive intersubjective reversal was not temporary but an acknowledgment of the common humanity, participants of cumulative and transformative social and ecological projects.

Z. Jiang
College of Economics and Management, Huazhong Agricultural University, 430070 Wuhan, China

H. Zhao
Guangxi Normal University Guilin, Guangxi 541001, China

F. Yun
Urban Planning and Design Institute of Shenzhen, Shenzhen 518000, Guangdong, China

W. Jiang (✉)
School of Economics and Business Administration of Central, China Normal University, 430079 Wuhan, China
e-mail: sgjyjiangjiang@163.com

D. Jones and A. J. Khan Chowdhury (eds.), *Environmental Resilience in the Pandemic Years 2020–2021*, Environmental Science and Engineering,
https://doi.org/10.1007/978-3-031-47757-7_11

Keywords Transformational tourism · Transformational festivals · Public and international travel · COVID-19

11.1 Introduction

The Covid-19 pandemic is changed every aspect of human life and its impact on the environment, comprising tourism. Government strategies and travel limitations in place in some countries are impacting perceptions of tourism during or after the pandemic is ended. After the Covid-19 pandemic ended, tourism is chosen as the preferred option. This reflects the view that travel today is fundamentally a human need (Wachyuni 2020). Several researchers investigated that when safety and health risks were seen as negative, peoples were likely to influence travelers' perceptions of destinations and seek them out with the quantity and quality of infrastructure consistent with necessary health protocols (Spenceley et al. 2021). The tourism industry can change short-term business strategies on the fly, prioritizing the health of tourists while considering economic risks. The tourism industry is estimated to recover after the year of the pandemic (Assaf 2020). Though there are risks facing, some tourists prefer a challenge that is more resistant to the various risks that can arise while traveling.

Many environmentally impacting amusement parks around the world are now reopening during the Covid-19 pandemic. According to interviews with environmental tourists, visitors feel that this is the entertainment they need, a fun experience that relieves stress, and that the waiting time at each attraction is short because they can play with satisfaction due to the reduced capacity.

During Covid-19 when tourism was stopped several alternative approaches were introduced to continue recreational activities such as responsible tourism and sponsors approach (Rowen 2020; Rubio-Mozos et al. 2020). As conceptual beginning factors or moral orientations for practice, those techniques are laudable and useful, whilst they method the bounds and conflicts of the entrepreneurial governmental economy. It changed into visible that in addition, a part of the hassle is the same old modes of difference among the areas and groups of visitors and hosts, which can obstruct efforts at joint action. Many of those issues were in addition masked with the aid of using the ephemerality and fragmentation of world enterprise practices, which effaces and displaces such issues as weather alternate as financial externalities, to be resolved with the aid of using future generations. Accompanying the prescriptive aforementioned approaches, "transformational tourism" was proposed to address the longstanding social and environmental challenges of tourism.

It became additionally investigated the transformative functions of a few festivals, particularly people who challenge social orders whilst urging prosocial behavior (Kwiatkowski 2020). This contains, amongst different things, dedication to involvement and accountability, the enjoyment of together conquering difficulties, and it includes pulling down or reversing social orders. Part of the confusion and venture in the use of phrases like "transformation" to impact alternate in enterprise and academia is the paradox of the time itself. Because one-of-a-kind streams of technological know-how originate their which means from one-of-a-kind genealogies. Many of the studies on "transformative tourism studying theory" indicates that it could be higher incorporated through a private transformation of values and behaviors to provide a clarification for how U.S. ladies go back to the hard work market and tour to a picnic spot (Jenkins 2003; Teoh et al. 2021).

The cutting-edge human traveler fortuitously embraces modifications in the journey and may be very able to be adapting to new surroundings, searching ahead to different matters even supposing they may be now no longer organized for it (Cohen 1972). Apart from the regulations on quarantine policies of greater than five months in a few countries. Most of the research reviewed suggested terrible mental consequences of quarantine, inclusive of signs of boredom, stress, anger, frustration, confusion, and economic loss (Brooks 2020). It is assumed that tourism revel in is essential to triumph over the terrible consequences. How tons of tourism revel in vacationer need for a selected excursion relies upon their tastes and favorites. However, now no longer tons of studies were carried out on tourism motivations throughout the pandemic.

The effect of Covid-19 on the tour motivations of Chinese tourists to tour overseas makes a specialty of the dialogue of stopping or rebooking global tours (Dedeoğlu and Boğan 2021). Therefore, this was predicted to function as a reference for vacation spot managers while formulating control techniques and accomplishing post-pandemic vacation spot improvement and might are expecting post-pandemic tourism trends. Further studies on tourism motivation are needed because it influences tourism calls for and traveler arrivals at endpoint (Wen et al. 2020). The Covid-19 pandemic has brought on an international tourism crisis. All events will paintings collectively to boost the transformation of tourism tailored to the modern situation. The tourism enterprise will use facts approximately findings associated with the motivations and styles of vacationers who tour post-pandemic to plot tourism sports and broaden regulations that assist entice greater vacationers to gain from tourism sports.

Tourism classification is frequently the problem of studies with the aid of using tourism researchers. Studies on traveler category intention to create organizations of vacationers or locations with comparable characteristics. Different kinds of vacationers have an impact on versions of tourism making plans with the aid of using accumulating tourism statistics from outside sources, locations, excursion durations, and itineraries earlier than making travel-associated decisions (Grzywacz 2016). In the age of connectivity, online social touch is emerged as a brand-new traveler habit, drastically influencing conduct and experiences (Buhalis 2019). Furthermore, said that it's miles essential to pick out traveler typologies primarily based totally on online and face-to-face social touch to formulate powerful advertising techniques for exclusive segments. Humans are commonly greater bendy with their surroundings and may extrude primarily based totally on environmental situations anywhere they are. Tourism awakens people's hobbies in normal life. When touch with the brand new makes them attractive, this appears to grow their willingness to depart their habitat and circulate quickly or maybe to another, fueled with the aid of using extended cognizance of the outdoor world.

Despite the newness of the tour enjoy, human beings are ill-prepared, sure with the aid of using ethnic cultures and beyond customs. There should be tailor-made to the desires of vacationers primarily based totally on man or woman options for novelty and familiarity with traveler destinations and became categorized into 4 one-of-a-kind types: The first one is prepared mass tourism (Cohen 1972). The kind of traveler who buys a tour package deal from a depended-on tour enterprise follows the tour enterprise's go-with-drift as deliberate with the aid of using the tour enterprise. It is famous however became a minimum novelty. 2nd was an independent mass traveler. The kind of traveler who buys a package deals with excursions however the excursion isn't always completely deliberate. Travelers have a few manipulate over their tour instances and plans and aren't tied to groups. However, all crucial preparations will nonetheless be made thru the tour enterprise. The itinerary isn't always a whole lot one-of-a-kind from the tour enterprise's plan. The awareness stage nonetheless leads, however the enjoy stage of novelty became grown. This kind of traveler arranges a unique experience that deviates from the standard route but presents snug and secure accommodation, transportation, and different traveler facilities. Trying to combo in with the brand-new environment of the traveled area, however, maintaining several the simple exercises and conveniences of the unique manner of life. These vacationers went out without tending to their simple desires. The fourth category is known as Wanderer. This kind of travel could be very one-of-a-kind for different vacationers who need precise tours, go to off-the-beaten-direction sites, submerge themselves in new environments, and stay just like the human beings of the locations they go to.

Additionally, considering the phenomena from the mindset of the so-called "mobility's paradigm," one can also don't forget that there are "many exceptional techniques one travels (e. g. communicatively, without a doubt and ingeniously) at some point of corporeal travel" (Lean 2012), and transformational gala aren't any exception. As the COVID-19 pandemic spread, local "way of life events" started to be lost across the world. Plans for the primary event, which were held yearly withinside the wasteland because 1990 and wherein price tag income cover a big part of the San Francisco-primarily based workplace staff's annual expenses, were additionally questioned. The businesses and groups they guide nonetheless face the impracticality of assembly in person, however, they go back to their roots and highlight participation, connection, and civic commitment withinside the face of quarantine.

In the early days of the COVID-19 disaster withinside the United States, face masks were a sensible requirement and a style object of innovative appearance for stills familiar with barren region conditions, however, they're dangerously in quick supply. Designers used advanced strategies to mass-produce homemade masks and prototyped 3-D printing era to reorganize and accelerate the manufacturing of ventilators to fill considerable health center shortages. Lockdowns, social separation, and worry of contagion are making humans worried (Moore and Lucas 2021) and depressed. People experience lonely and tired of their isolated way of life (Pedersen et al. 2012). Travelers subject to government or voluntary travel restrictions (for fear of contagion) experience an intrinsic motivation to travel, breaking the monotony of social distancing and isolation.

Social distancing existence was making vacationers angry, depressed, and bored giving manner to new journey tendencies (Roy and Sharma 2021). Travelers need quick journeys to interrupt the monotony of quarantine life. Travelers are seeking out holidays and are increasing rotating to the net for evaluations and suggestions (Parra-López et al. 2012). Smart the cap potential to effortlessly get admission to vlog posts and evaluations on their hand-held gadgets spawned new journey tendencies along with Video blogs (vlogs) have emerged as a famous manner of communication. Research indicates that 60% of vacationers determine their subsequent vacation spot after looking at a journey video. These journey vlogs result in journey motivation. Online evaluations of studies and locations inspire humans to journey. The outcomes had been primarily based totally on the maximum number of respondents, i.e., 85% stated they had been making plans to journey after the pandemic, 20% stated 'maybe' and the most effective four% stated 'no'. The preference of journey locations changed into greater withinside the path of home locations. Concerning the environmental effect of COVID-19, the maximum favored sort of tourism is nature tourism (66%), observed through normal tourism (19%) and cultural tourism (8%). The deliberate journey duration changed into ruled through quick-time period periods of one to four days. Survey statistics on journey goals are proven in Table 11.1.

Table 11.1 Survey statistics to travel responses

Questions item	Options	Answer %
traveled after pandemic period	Yes	85
	Maybe	20
	No	4
Desired destinations and their effects	Local	60
	International	5
	Local and international	45
Type of tourist places	Nature or festivals tourism	66
	Religious tourism	6
	Cultural tourism	8
	Culinary tourism	19
	Others	4
Duration of tours and recreational activities	1–3 days	65
	4–8 days	45
	> 8 days	20

11.1.1 Conceptual Framework

Previous studies explored the social, cultural, personal, and mental elements that were an impact on push and pull reasons, which in addition was an impact on journey selections (Jang and Cai 2002). The pandemic (COVID-19) outstanding instances make it important to discover the reasons to explain for the journey selection to close places as lengthy-distance journey remains limited (Fig. 11.1). Through investigations, the look identifies the frenzy and pull reasons and finds the important thing elements that hinder journey plans throughout those instances of Covid-19. More significantly, pull elements weren't sufficient electricity except the ones precipitated via way of means of push elements affecting journey selections throughout the pandemic time (Abdullah et al. 2021). Travelers' inner motives push them to go to close vacation places throughout the pandemic time (Guillet et al. 2011). This reality motivates them to check the speculation that the social-mental elements that push humans to journey for a day-lengthy excursion throughout COVID-19 have a wonderful impact on the journey aim (pull elements) to a selected destination. The threat belief of travelers will become a critical issue in influencing behavioral aims throughout the pandemic time (Bae and Chang 2020).

Physiological inspiration is for relaxation, fitness, comfort, collaborating in sports activities, and recreation. This inspiration comes from inside a person's frame who desires to experience relaxation, lessens worrying muscle mass from running or reading tough, and revives a pressured mind, so it's far feared that it'll influence bodily and mental fitness if now no longer done. Social motivation or relational motivation is the choice to socialize, meet different humans to speak approximately new matters, go to friends and family, meet partners, and do matters which are taken into consideration to be pleased with themselves with what's performed or an honor (Dwisaputra 2017). Another factor is the want to find out about the journey for social

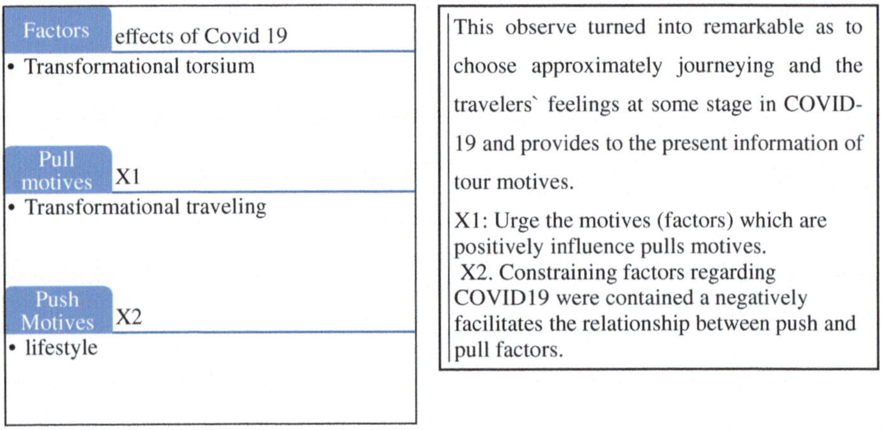

Fig. 11.1 Conceptual framework for transformation

dealings, intermingling with nearby folks who can provide travelers with studying experiences (Pearce and Lee 2005). This examines labels the contemporary worldwide issues, forms of travelers, and journey motivations after the Covid-19 pandemic. The objectives of this paper are to study modifications in tourism phenomena, recognize modifications in journey traits after the Covid-19 pandemic, and decide travelers' motives withinside the perspective of travel categorization.

11.2 Methodology

Tables 11.2 and 11.3 display that the Covid-19 pandemic is converting all components of life, along with tourism. This is associated with the converting attitude of vacationers approximately a way to cope with the Covid-19 pandemic. A questionnaire survey approach was implemented. This study approach is certainly considered one among much equipment utilized by researchers withinside the tourism sector. Data series changed into achieved via way of means of dispensing a web questionnaire (Google Forms) through WhatsApp airing messages. An online survey was changed into performed to assist acquire reactions from special components of China. Questionnaire surveys are true for collecting statistics approximately character traits, perceptions, attitudes, and behavior. The levels of a survey are survey design, survey method choice (in this example online survey), and respondent choice (Balcombe et al. 2009; Clifford et al. 2016).

We observed that the visitor attitudes (effective, terrible, impartial) were related to the rankings in their journey enjoy at the day. Sentiment evaluation (Table 11.2) suggests that the maximum opinions (74.68%) reflected effective excursion sentiment at some stage in his one-day excursion at some stage in a lockdown. On the alternative hand, 21.41% of opinions have been terrible and 4.91% of opinions have been impartial approximately the timing of the excursion pandemic. Interestingly, 54.14% of the opinions specified mild to very effective emotions toward the ride because the tourists were bored at domestic because of the extended quarantine. They

Table 11.2 Sentiment analysis of travelers' and transformation tourism

Valence	Degree of valence*	Number of comments	Percentage
Positive		385	74.68
Highly	Above 0.7	123	24.78
Medium positive	0.4<, < 0.7	155	30.64
Low negative	0.1<, < 0.4	98	19.5
Highly negative	Below/0.7	109	21.41
Medium negative	(– 0.7 </< – 0.4)	42	8.15
Low negative	(– 0.4 </< – 0.1)	9	1.6
Neutral	Nearly zero (– 0.1 </< + 0.1)	25	5.0

Table 11.3 Differentiate the variable and percentage

F-test sample analysis		
	Variable 1	Variable 2
Mean	118	22.955
Variance	14,207.71429	521.1351143
Observations	8	8
Df	7	7
F	27.26301471	
P(F ≤ f) one-tail	0.000144738	
F critical one-tail	3.78704354	

loved a quick ruin withinside the shape of a quick excursion to a close-by beach. The excessive percent of effective sentiment was attributed to the lively function of the virtual influencer in instilling belief and self-assurance in their fans through "sharing situational understanding" at some stage in the pandemic (Lo and Lee 2011; Do et al. 2022). The video blogger shapes belief and self-assurance in a YouTube put up on "shared situational awareness" approximately full-day excursions with the aid of using sharing how tourists can revel in a quick excursion even as preserving protection and hygiene at some stage in the pandemic (Do et al. 2022). This belief and integrity are in addition reinforced with the aid of using different vlog fans to contribute to their journey reviews and assisting their vlog reviews.

The findings verify the significance of virtual influencers constructing effective sentiment amongst customers with the aid of using presenting fact-primarily based data in shared social networks. Emotional strain at some stage in a plague wreaks havoc on people's minds. People crave quick breaks to eliminate the depression, tension, and tedium of a hikikomori life. To higher recognize the visitor's emotional state, the consequences of the sentiment dictionary-primarily based evaluation of the whole score dataset are proven in Table 11.3. As a result, the primary emotion showed that the withinside the evaluation is effective ($\alpha = 0.2737$), accompanied with the aid of using worry ($\alpha = 0.1604$) and joy ($\alpha = 0.1224$). Therefore, this evaluation confirmed tourists' blended emotions approximately same-day journey at some stage in the pandemic. Additionally, feelings withinside the shape of worry and tension also are not unusual place area among tourists and their carriers of hospitality services (Sönmez et al. 2020). A purchaser fears their COVID-19 infection, even as provider issuer faces worry and tension approximately.

Table 11.4 suggests life and human feelings at some stage in the pandemic. During COVID-19, tourists' feelings can happen withinside the shape of worry, anxiety, and sadness. These emotions can result in the formation of poor emotions approximately tour at some stage in the pandemic. These effects additionally discovered full-size emotional sharing of worry ($\alpha = 0.1605$), sadness ($\alpha = 0.0981$), anger ($\alpha = 0.0078$), and typical unfavorable consequences on tourism ($\alpha = 0.1123$). However, considering withinside the 'shared situational awareness' was produced with the aid of using bloggers and their fans became related to high-quality conduct and effectiveness (α

Table 11.4 Lifestyle and human emotion association lexicon analysis

Emotional effects	Frequency	Output (α)
Positive action	580	0.273
Anger act	160	0.0078
Disgust stress	20	0.0089
Trust	12	0.0005
Fear	330	0.1605
Happiness	260	0.123
Negative effect	238	0.1123
Sadness	200	0.0981
Surprise	10	0.0495

**Variables* Positive, anger, disgust, trust, fear, Happiness, negative, sadness, and surprise, represent the types of emotional effects of states

= 0.273) and luck (α = 0.123), however few at some stage in the COVID-19 period. possibilities and excursion plans. These high-quality emotional states in vacationers might also additionally cause a preference for brief excursions whilst following protection protocols.

11.3 Results and Discussion

The consequences of information evaluation are displayed with related moods. However, as proven in Table 11.5, best often taking place phrases with related moods. Data evaluation confirmed that a few phrases including travel, 'tour', 'beach', and 'experience' were greater often related to the wonderful sentiment (primarily based totally on time period frequency) than with terrible or impartial sentiment. found out to be high. We additionally observed that phrases like "lockdown," "safety," "infection," and "hygiene" had been related to terrible feelings in place of wonderful or impartial feelings. As such, those are worries concerning travel-associated selections for the duration of the pandemic.

Based on phrase frequency, the pinnacle 3 key phrases related to high-quality feelings were tour (155), seashore area (151), and enjoyed (Table 11.6), at the same time as the pinnacle 3 key phrases related to poor feelings where Keywords are lockdown (80), safe (45)., and infections (70). It turned into discovered that only a few key phrases have been sentiment impartial. The pinnacle 3 phrases for impartial temper are hotel (78), travel (64), and picnic (50). We additionally discovered that phrases such as 'time', 'stress', and 'coronavirus' have full-size impacts, each high-quality and poor, on travelers' studies and plans for the duration of the pandemic. Few people (20%) confirmed greater poor feelings than high-quality or impartial feelings.

Table 11.5 The statically and ANOVA result

SUMMARY	Count	Sum	Average	Variance
Row 1	2	580.273	290.1365	168,041.7
Row 2	2	160.0078	80.0039	12,798.75
Row 3	2	20.0089	10.00445	199.822
Row 4	2	12.0005	6.00025	71.994
Row 5	2	330.1605	165.0803	54,397.05
Row 6	2	260.123	130.0615	33,768.03
Row 7	2	238.1123	119.0562	28,295.28
Row 8	2	200.0981	100.0491	19,980.38
Row 9	2	10.0495	5.02475	49.50623
Column 1	9	1810	201.1111	33,972.11
Column 2	9	0.8336	0.092622	0.007879

ANOVA

Source of variation	SS	df	MS	F	P-value	F crit.
Rows	136,012.3891	8	17,001.55	1.001825	0.499003	3.438101
Columns	181,837.9479	1	181,837.9	10.7149	0.011299	5.317655
Error	135,764.5628	8	16,970.57			
Total	453,614.8999	17				

Figure 11.2 suggested the transformation of attachment and tourism during the Covid 19, the effect of a journey and the lives of these changed, and what this must appear like. Overall facts or frequency plots (improved facts evaluation in Fig. 11.2) confirmed that the pinnacle 3 phrases had been lockdown (80), tour (155), and seashore area (151). Of the pinnacle spots, the handiest phrases, 'lockdown' and 'safe', have greater bad than fantastic sentiments. The consequences display that tourists have greater fantastic attitudes toward the journey than bad facts, as they favor eliminating an uneventful life (x = 10) and uninteresting life (x = 11).

The frequencies and related sentiments (see Table 11.3) monitor the intrinsic and extrinsic elements that govern patron minds concerning taking same-day excursions throughout those extraordinary times. Data suggests that bloggers are in a country of intellectual alertness (push motivation) and are intrinsically motivated. The few phrases that dictate the phrase pool push signs are enjoying, experience, life, relax, and refresh (see Table 11.4). Bloggers are mentally excited to participate in a brief day experience to relax, refresh, and feature an exciting experience.

It is the result of a hard and fast of things and motivations. This presents a trademark of what motivates travelers to take day trips. The element that dominates the listing is the 'seashore area'. Other elements that appeal to tourists to a vacation spot are seashore cleanliness (cleanliness), excursions, cleanliness, and motel accommodations. Travelers select now no longer to do so, even though they're mentally agitated and sense the urge to tour. According to a survey (statistica.com), 35% of

Table 11.6 Sentiments and traveling

Positive review	Frequency	Negative review	Frequency	Neutral review	Frequency
Beach area	151	Beach area	15	Beach area	10
Tour	155	Tour	25	Tour	3
Enjoy	121	Enjoy	0	Enjoy	0
Lockdown	80	Lockdown	32	Lockdown	4
Safety	45	Safety	123	Safety	2
Clean	80	Clean	0	Clean	1
Experience	5	Experience	90	Experience	4
Hotel	78	Hotel	9	Hotel	0
Scenic	67	Scenic	70	Scenic	9
Lifestyle	60	Lifestyle	1	Lifestyle	2
Trip	64	Trip	15	Trip	1
Hygiene	59	Hygiene	1	Hygiene	8
Beautiful	11	Beautiful	0	Beautiful	1
Picnic	50	Picnic	58	Picnic	0
Relaxation	55	Relaxation	0	Relaxation	9
Journey	49	Journey	2	Journey	1
Time	41	Time	35	Time	2
Depression	35	Depression	2	Depression	0
Refreshing	28	Refreshing	15	Refreshing	0
Stress	36	Stress	30	Stress	0
Coronavirus	35	Coronavirus	2	Coronavirus	0
Transport	30	Transport	0	Transport	0

Indian tourists say they don't have any tour plans and could make selections after the outbreak develops. So, it turns into thrilling to perceive the complete adventure and transformation (Table 11.7).

During the COVID-19 outbreak, boundaries were diagnosed that averted aspiring tourists from traveling. NLP evaluation as provided in Table 11.7 indicated that "lockdown" turned into the most important inhibitor. This result additionally indicates that there are worries about 'infection', 'safety' and 'hygiene' in visitor regions as different impediments. That recognized the exceptional metrics for every motivation for excursion decisions (Roy and Sharma 2021). However, on this examination, perception is incomplete whilst those phrases aren't related to signs of deeper perception. Tourism has come to be one of the maximum affected sectors withinside the world. There are, however, a few optimisms as visitors' urge for the food stays strong. To confirmed that almost all of the respondents stated they turned to make plans for an excursion soon, 0–6 months after the give up of the Covid-19 pandemic (Wachyuni 2020).That expected the tourism enterprise could get better after his yr withinside the pandemic (Assaf 2020). Like numerous countries, which includes China, tourism

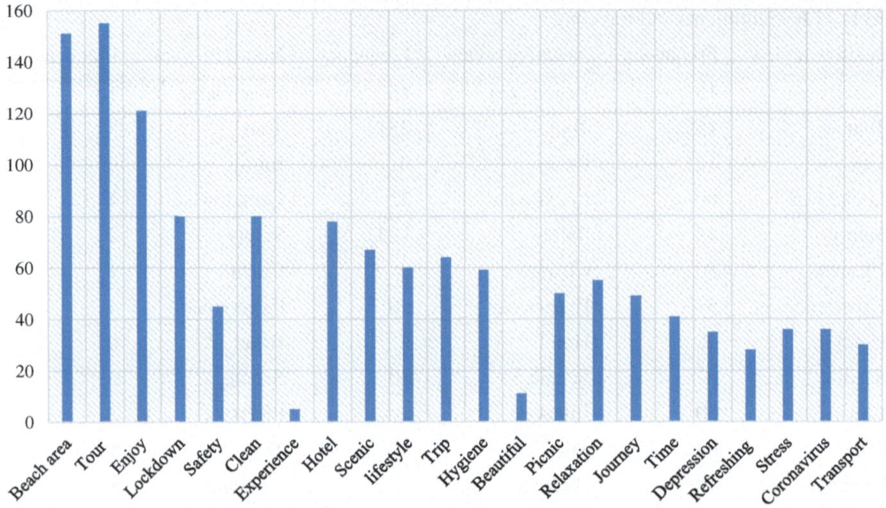

Fig. 11.2 The transformation of environmental tourism and traveling impact due to covid-19

Table 11.7 Inhabitation factor of transformation frequency (negative sentiment)

Inhibiting factors (term)	Frequency (negative sentiment)
Lockdown	115
Safety act	90
Infection	70
Hygiene	60
Coronavirus	30
Social	25

hobbies withinside the United States of America started in June 2020, adapting to strict fitness protocols at same time as tracking the evolving situation, which was associated with Motivation is a hard and fast organic and mental wish (Crompton 1997). Motivation is consequently a crucial issue withinside the decision-making process (Smith et al. 2010).

As a result, we located that tourism motivation is ruled via way of means of bodily or physiological motivation. Boredom for the duration of self-quarantine for the duration of the pandemic encourages respondents to relax, bodily and mentally rest and journey proceeding to make the frame extra comfortable and comfortable Nature tourism locations can be appropriate to satisfy the rest wishes of examining subjects (Kusumaningrum and Wachyuni 2020). This locating is steady with numerous research of visitor motivations for touring herbal visitor sites. recognized her 5 motivating elements for traveling to South African maritime locations, which include studies and evidence (Richards and Morrill 2020). This consists of get-away and rest, vacation spot attraction, socializing, private connections, region attributes, and journey characteristics. In addition, the fable reason is her 0.33 respondent's choice. According to fable motivation is the traveler's choice to get away from the monotony of ordinary lifestyles and is associated with mental pride (Ory and Mokhtarian 2005).

In a routine of each day lifestyle with extreme social regulations for everyone, respondents need to journey for non-secular pride and comfort from boredom. Cultural motivations, on the opposite hand, come finally in respondents' choices. This motivation is steady with the studies (Wachyuni 2020) that cultural tourism isn't a prime visitor vacation spot withinside the post-pandemic era. However, the respondent prefers nature tourism and gastronomy. This case arises due to the fact respondents favor to relax, permit pass of boredom, and featuring new reviews to meet their journey pride.

11.4 Conclusions

This study affords theoretical consequences that the Covid-19 Pandemic altered the fashion of mass tourism into non-public tourism. Also, this observation affords up-to-date records associated with traveler motivation after the Covid-19 Pandemic. Above and beyond the theoretical implications, this study additionally affords sensible suggestions that vacation spot managers want to modify the characteristics, typology, and enthusiasm of tourists.

This research additionally affords to enter for journey marketers and vacation spot managers to modify to technology. Other studies that may be accomplished in addition is to pick out what herbal tourism sports are the vacation spot selections, for instance, mountaineering mountains which can be typically executed during the Pandemic in China.

It was observed that they supplied the info on push, pull, and inhibiting elements during the pandemic time withinside the Indian perspective and centered on a particular kind of vacation spot (beach). The textual content mining technique won't offer in-intensity perception into the reviewers' socio-financial info in the Vlog posts, not like Trip Advisor affords a few demographic info of the reviewers on its platform and all maximum each kind of network laid low with COVID-19.

Future research can also additionally carry out survey-primarily totally based strategies and examine fashions primarily based totally on responses from extraordinary sociodemographic subgroups. Survey effects are greater complete with qualitative data, along with in-intensity interviews with respondents.

References

Abdullah M, Ali N, Hussain SA, Aslam AB, Javid MA (2021) Measuring changes in travel behavior pattern due to COVID-19 in a developing country: a case study of Pakistan. Transp Policy 108:21–33

Assaf A, Scuderi R (2020) COVID-19 and the recovery of the tourism industry. Tourism Econ 26(5):731–733. https://doi.org/10.1177/1354816620933712

Bae SY, Chang PJ (2020) The effect of coronavirus disease19 (COVID-19) risk perception on behavioural intention towards 'untact' tourism in South Korea during the first wave of the pandemic (March 2020). Curr Issues Tourism: 1–19.https://doi.org/10.1080/13683500.2020.1798895

Balcombe K, Fraser I, Harris L (2009) Consumer willingness to pay for in-flight service and comfort levels: a choice experiment. J Air Transp Manage 15(5):221–226

Buhalis D, Sinarta Y (2019) Real-time co-creation and nowness service: lessons from tourism and hospitality. J Travel Tourism Market 36(5):563–582

Brooks S, Smith L, Webster R (2020) The psychological impact of quarantine and how to reduce it: rapid review of the evidence. Lancet 395(10227):912–920. https://doi.org/10.1016/S0140-6736(20)30460-8

Clifford N, Cope M, Gillespie T, French S (2016) Key methods in geography, 3rd edn. Sage Publication, Glasgow

Cohen E (1972) Toward a sociology of international tourism. Political Econ 39(1):164–182. Retrieved from http://www.jstor.org/stable/40970087

Crompton J, McKay S (1997) Motives of visitors attending festival events. Ann Tourism Res 24(2):425–439. https://doi.org/10.1016/S0160-7383(97)80010-2

Dedeoğlu BB, Boğan E (2021) The motivations of visiting upscale restaurants during the COVID-19 pandemic: the role of risk perception and trust in government. Int J Hospital Manage 95:102905

Do B, Nguyen N, D'Souza C, Bui HD, Nguyen TNH (2022) Strategic responses to COVID-19: the case of tour operators in Vietnam. Tourism Hospital Res 22(1):5–17

Dwisaputra M, Achnes S (2017) Motivasi Pengunjung ke Kota Wisata Bukittinggi Sumatera Barat. Jurnal Online Mahasiswa Fakultas Ilmu Sosial Dan Ilmu Politik Universitas Riau OM FISIP 4(2):1–13

Grzywacz R, Zeglen P (2016) Typology of tourists and their satisfaction level. Sci Rev Phys Culture 6(1):5–16

Guillet BD, Lee A, Law R, Leung R (2011) Factors affecting outbound tourists' destination choice: the case of Hong Kong. J Travel Tourism Marketing 28(5):556–566

Jenkins O (2003) Photography and travel brochures: the circle of representation. Tour Geogr 5(3):305–328

Jang S, Cai LA (2002) Travel motivations and destination choice: a study of British outbound market. J Travel Tourism Marketing 13(3):111–133

Kwiatkowski G, Oklevik O, Hjalager AM, Maristuen H (2020) The assemblers of rural festivals: organizers, visitors and locals. Euro Plann Stud 28(2):255–272

Kusumaningrum DA, Wachyuni SS (2020) Promo cashback GoPay Terhadap Minat Beli bubble drink (Studi Kasus: Yu Cha Indonesia, Pluit). J Tourism Econ 3(1):23–30

Lean GL (2012) Transformative travel: a mobilities perspective. Tourist Stud 12(2):151–172

Lo AS, Lee CY (2011) Motivations and perceived value of volunteer tourists from Hong Kong. Tourism Manage 32(2):326–334

Moore KA, Lucas JJ (2021) COVID-19 distress and worries: the role of attitudes, social support, and positive coping during social isolation. Psychol Psychother Theor Res Pract 94(2):365–370

Ory DT, Mokhtarian PL (2005) When is getting there half the fun? Modeling the liking for travel. Transp Res Part A Policy Pract 39(2–3):97–123

Parra-López E, Gutiérrez-Taño D, Diaz-Armas RJ, Bulchand-Gidumal J (2012) Travellers 2.0: motivation, opportunity and ability to use social media. In: Social media in travel, tourism and hospitality: theory, practice and cases, 171–187

Pearce PL, Lee UI (2005) Developing the travel career approach to tourist motivation. J Travel Res 43(3):226–237
Pedersen PV, Andersen PT, Curtis T (2012) Social relations and experiences of social isolation among socially marginalized people. J Soc Personal Relationships 29(6):839–858
Richards G, Morrill W (2020) Motivations of global Millennial travelers. Revista Brasileira De Pesquisa Em Turismo 14:126–139
Roy G, Sharma S (2021) Analyzing one-day tour trends during COVID-19 disruption–applying push and pull theory and text mining approach. Tourism Recreation Res 46(2):288–303
Rowen I (2020) The transformational festival as a subversive toolbox for a transformed tourism: lessons from burning man for a COVID-19 world. Tourism Geogr 22(3):695–702
Rubio-Mozos E, García-Muiña FE, Fuentes-Moraleda L (2020) Sustainable strategic management model for hotel companies: a multi-stakeholder proposal to "walk the talk" toward SDGS. Sustainability 12(20):8652
Sönmez S, Apostolopoulos Y, Lemke MK, Hsieh YCJ (2020) Understanding the effects of COVID-19 on the health and safety of immigrant hospitality workers in the United States. Tourism Manage Perspect 35:100717
Smith S, Costello C, Muenchen RA (2010) Influence of push and pull motivations on satisfaction and behavioral intentions within a culinary tourism event. J Quality Assur Hospital Tourism 11(1):17–35
Spenceley A, McCool S, Newsome D, Báez A, Barborak JR, Blye CJ, Zschiegner AK (2021) Tourism in protected and conserved areas amid the COVID-19 pandemic. Parks 27:103–118
Teoh MW, Wang Y, Kwek A (2021) Conceptualising co-created transformative tourism experiences: a systematic narrative review. J Hospital Tourism Manage 47:176–189
Wachyuni SS, Kusumaningrum DA (2020) The effect of COVID-19 pandemic: How are the future tourist behavior. J Educ Soc Behav Sci 33(4):67–76
Wen J, Kozak M, Yang S, Liu F (2020) COVID-19: potential effects on Chinese citizens' lifestyle and travel. Tourism Rev 76(1):74–87

The manufacturer's authorised representative in the EU is Springer
Nature Customer Service Centre GmbH, Europaplatz 3, 69115 Heidelberg,
Germany. If you have any concerns regarding our products, please
contact ProductSafety@springernature.com

Printed and bound by CPI Group (UK) Ltd, Croydon, CR0 4YY
30/09/2026
02225600-0003